iOS App Development for Teens by Teens: Beginning iOS App Development
by Randall (a.k.a. Chip) Beck II

ISBN-13 (printed): 978-0-9916256-0-4

AF777479

ISBN-13 (ePub): 978-0-9916256-1-1

ISBN-13 (PDF): 978-0-9916256-2-8

Printed in the United States of America

Published by Ch1pa Books

funded with
KICK
STARTER

Chip's one and only selfie

About the Author

Chip Beck, a high school sophomore at the time of writing, wrote his first app in between his seventh and eighth grade year. His first app, a very basic calculator app, was published on the App Store when he was only thirteen. The amazing feeling of being published on the App Store encouraged Chip to continue with his app development.

Through the years to come, Chip developed a total of fifteen apps independently while he founded Ch1pa Software and still went through a normal school life. When asked what kind of apps he makes, he jokes with a simple "Whatever comes to mind!", and then shows his questioner a few apps that range from an "anti-joke" generator to an app developed in Simplified Chinese that incorporates a variety of Chinese social networks and messaging services.

Chip lives in a small town in northwestern Pennsylvania with his parents and sister (who is away at college). Whenever Chip isn't coding or going to school, he is playing his saxophone and piano, playing tennis, performing in a musical, or sleeping.

For any questions, comments, or concerns, email Chip at ChipBeck@Ch1paSoft.com.

Why Write This Book?

The question that is always asked, over and over by friends, teachers, and family members. I can answer this one question with two different answers, so I will. One of the two main reasons for writing this is to share the knowledge, and make it simple. When I became interested in programming, I purchased a few books claiming that they were for "beginners", they were not. These books had assumed that you already knew how to program, and understood what their code had done or meant. As someone who had just come into the world of programming, I did not understand any programming concepts and was plain confused. I felt the need to write a book where a true beginner could learn how to make an iOS app without prior programming knowledge. I wanted to let people know that ANYBODY can make an app! The other big reason I wanted to write this book is because I can! When people ask me "Why?", I respond with "Why not?". I live in a country where you are free to do what you please. I live in a time where anyone of any age can sit down and write a book on what they love, and self publish it. Why not reap the benefits? Why not break the boundaries of what people think a stereotypical teen can do? You are what you believe you are, why not join me?

Chip Beck, 15

PREREQUISITES

I hate that there have to be any prerequisites at all, but for this book you must have a computer running Mac OS X 10.8 Mountain Lion or later with at least a temporary Internet connection. I am glad to say however that owning a Mac is the only prerequisite that is required in following the tutorials in this book. If you don't have Mac or are not running OS X 10.8 or later, refer to the list below which contains all of the Macs compatible with OS X 10.8.

Mac (Mid 2007 or newer)

MacBook (Late 2008 Aluminum, or Early 2009 or newer)

MacBook Pro (Mid/Late 2007 or newer)

MacBook Air (Late 2008 or newer)

Mac mini (Early 2009 or newer)

Mac Pro (Early 2008 or newer)

Xserve (Early 2009)

*Note: you may attempt to use a Windows computer running Mac OS X 10.8 using VirtualBox, but I can't guarantee any results.

A recommendation that I have is that you own or have access to a device with iOS 7 or later, such as an iPhone or iPad, to run the apps we create in the book. This is not required as all apps we make are compatible with the iOS Simulator that comes with Xcode, the free app provided by Apple that we will make apps in. Downloading and installing Xcode will be explained in the book.

iOS App Development for Teens by Teens Financial Supporters

From January 25th, 2014 through February 22nd, 2014, I ran a Kickstarter project to raise enough money to publish this book. I received overwhelming support from peoples of all nations. In total, the project raised $2,207 and distributed this book to sixteen countries!

Here are some of the people who helped me out along the way:

Jeanette Anderson
Chris Anderson
Sandra Armstrong
Robert Armstrong
Fabio Bargero
Christina Blackburn
Andreas Bøe
Connie Carey
Guyver Designs
Yvonne English
Kyle English
Larry Enoksen
Paula Ference
Trevor Ference
Spencer Ference
Chuck Ference
AC Fuller
Giovanni
Jeff Heaton
Barbara Hild
Maurine Hobbs
William James Hobbs
B. W. Hulette
Laura Jackson
Jeff Jackson
Tom Jackson
Jeffrey Jackson
Shawn "phChun" Jordan
Paula Kalp
Wayne Kalp
Jessicalind S Ah Kit
Carl L Gilchrist
Whittnee LaChapell
P. A. Lavery
Tax Mam, Inc.
Jonathan Marrie
James Muschette, III
James Nevison

Steve Palumbo

Gregory Rapley

Randolph "D. Psyko" Richards

Rudare Communications, SL

Karl Schmidt

Kjerstin Schroeder

Stephen Shellenberger

Roelof Siepel

Todd Smith

Madge Tamber

Ron Tamber

The Rev. Adam Trambley

TRG International

Chuck Truxal

Judy Truxal

Joe Walker

Richard G Wiater

Susan Yount

Rick Yvanovich

I would like to thank Mrs. Jeannette Anderson, the one who got me interested in computer science and programming, and would also like to thank Mr. David Gibb, Mrs. Alissa Sgro, Mrs. Cheri Manners, Mrs. Nicole Porter, Dr. David Lineman, Mr. William Dzuricsko, Mr. Steve Garay, Mrs. Julie Freeman Morris, my principals, my friends, my Kickstarter backers, and most of all, my family, for all of their support. I'd like to especially thank my mother, Jacque Beck, for proofreading the grammar of the book, and my father, Randy Beck, for printing the book.

To Steven Paul Jobs, my inspiration for writing.

-Chip Beck

Chapter 1

Introduction to Computer Programming

As defined by the New Oxford American Dictionary, a computer programmer is "a person who writes programs for the operation of computers, especially as an occupation". The purpose of this book is to make you, the teenage reader, a young computer programmer. As referenced by many older people as a "pocket computer", the smartphone has become one of the most sophisticated devices owned by the average consumer in human history. All together, calling a smartphone a "pocket computer" is an understatement. The smartphone is a media center. The smartphone is an internet browser. The smartphone is a casual game console. The smartphone is of course, a phone. When I think of computer programming, I think of all the marvelous programs and apps I use everyday that make my life more enjoyable.

Although this book is focused on teaching you the basics of iOS app development, we must first go through the basics of programming. This will include the structure, logic, syntax, naming and commenting, and a short "Hello World!" program in the programming language called Ruby. If you are already familiar with what programming is and what it looks like, feel free to skip onto Chapter 2, where the beginning of iOS app development content begins, or get a refresher by continuing from here.

1.1 THE STRUCTURE OF CODE

Every programming language has it's own structure that you must code in. In this chapter we will be focusing on the language, Ruby, which is a good language to learn about the basics of programming as it has a very simple structure.

Most programming languages read code from left to right, then top to bottom, just like the English language. The version of Ruby we will use is Terminal based which means that it is less object oriented (OOP) and should run straight through with the exception of loops which I will explain in a couple sections.

An example of the structure of code I have presented below should demonstrate what basic code looks like. The code is in Objective-C, the language of iOS apps, which we will get into the next chapter.

```
#import <UIKit/UIKit.h>
int main(void) {

NSLog("Hello World");

}
```

Let's analyze the code!

Line 1's *#import* is called the directive and *UIKit/UIKit.h* is the header. These will be explained in a later section. Line 2 starts with *int main(void)* which starts the "main" part of the program which is where all the fun happens! Line 4 is difficult to explain at the time and will be explained in a later chapter. All I can say is that it outputs the text *Hello World* into the console. The curly brackets ({}) are where the real content of the program lies, but again, all of this and more will be explained in the sections to come in this chapter.

Whenever it comes to console programming in C++, the placement and order of code is very important, as the output is different based on the order the code is placed. Always keep this in mind whenever you are programming.

1.2 Data Types and Comments

Data types are to programs as sock brands are to jocks, very important. In programming, data must be categorized in order for commands to read the data properly. Also by categorizing data, you are able to conserve memory and storage space. In simple terms, data types store "stuff". They can store text, numbers, yes or noes, lists, and much more. The chart below explains the different types of data and what they are able to store.

Chart of Basic Data Types

	Description	Size	Example
char	One character	1 byte	'L'
string	Combination of characters that can be of any type (math functions don't work with strings)	4 bytes + length in characters of string	"Who are you?"

	Description	Size	Example
int	A whole number (positive or negative)	4 bytes	143656
float	A floating point number (has a decimal point)	4 bytes	3.14
double	A very large number with a floating point	8 bytes	1.1E+16
short	A short integer (between -32768 and 32767)	2 bytes	-145
long	An integer up to 64 bits	4 bytes	345238

	Description	Size	Example
Boolean (BOOL)	True or false statement (1 or 0)	1 byte	TRUE

Data types are typically declared by the programmer on an as needed basis at the beginning of program. Variables are all declared in a very general way. In most languages, they are declared in the order of *type*, *assigned name*, then *value*. For example:

```
int example = 54;
```

In the case of declaring multiple variables of the same type on the same line, you can simply add a comma between assigned names. For example:

```
int example, sample, synonym;
```

Then you can assign values as such:

```
example = 43;
sample = 2;
synonym = 90;
```

All other number data types are declared in the same fashion as ints.

Declared chars are given a value with apostrophes such as:

```
char example = 'A';
```

Declared strings are given a value with quotation marks such as:

```
string example = "iOS App Development for Teens
by Teens is awesome!";
```

Comments

Whenever you need to make notes or need to temporarily remove code from a program, comments are there to help. Comments can be little notes to the side of code that remind you of something, or they can be put in front of code in order to nullify it. A comment is not read by the computer whenever it goes down the code, so you can type whatever you want, or test out different methods of doing a task without having to delete the code. There are two types of commenting in code. Line comments only affect the part of the line past the location they are placed, and are typed as two forward slashes (//). Below is an example of code followed by a line comment.

```
int example = 4; //Sample of comments
```

The other type of comment is a block comment which can be as long and wide as you make them. Block comments are created by typing a forward slash followed by an asterisk *(/*)*. The block comment is ended by typing an asterisk followed by a forward slash *(*/)*. Anything placed in-between the two parts of the block are part of the comment and are not read by the program. Below is an example of a block comment.

```
/*
     int example = 4; is not being read!
     Neither is this!
     Or this!
*/
```

1.3 Understanding Programming Through Similes and Metaphors

A program, as defined by the Merriam-Webster Dictionary, is "a sequence of coded instructions that can be inserted into a mechanism". A program is simply a set of instructions, provided by us, that tells the device to do "something". That "something" could be creating a grocery list, a "hot or not" manager, a quote generator, a homework solver, or ANYTHING that you can dream of. With programming, there are unlimited possibilities!

I know from personal experience that programming is very confusing in the beginning, but the confusion can be cured through relating programming to simple things from the real world. As you may have read this section's title, this section is all about getting to know programming better through relations. Below are some of my favorite programming metaphors and similes.

List of Programming Metaphors and Similes

- You and I are in a car. Tell me exactly how to drive to the destination without getting in an accident.
- Coding is telling a person without self-thought how to make a peanut butter and jelly sandwich.
- Programming is like building with Legos, you stack small pieces together to make bigger things. Not every piece can be stacked. You only have a limited number of Legos. Stepping on one "hurts like Hell", but playing with them are fun.
- Programming is like raising a child, no matter how many books you have read about it, you still feel like you have no idea what you are doing, and you still love it in the end.

A very important concept in programming is understanding the structure and layout of code; there are metaphors for that too!

- Programming is like a game plan, you have a foundation, an idea of what you want to accomplish, and players you need to tell what to do.

- The layout of code is a structure, a building that starts with a foundation and blueprint that is built upon with supports, rooms, stairs, and last but not least, the aesthetic design.

As you can probably tell by now, I love metaphors and similes! I believe that they are vital in understanding a new and unfamiliar concept. They give the reader or listener a chance to pick up a concept by simply relating or comparing it to something they already know and are comfortable with.

I feel it is now time for us to code our first program together!

1.4 Coding Our First "Hello World" (In Ruby)

So far in this chapter, we have learned about the basic structures and principles of coding, how it works, and how it is read by the computer. We have covered the basic types of data found in nearly every object oriented programming (OOP) language, and have gone through some extensive programming metaphors and similes.

What we will be doing, is using the built in Terminal application to code in the Ruby programming language and output the message "Hello World!". By completing your first Hello World, you are automatically inducted into the pretend consortium of programmers, and must fulfill your duty of learning about iOS app development throughout the rest of the book.

Getting Ready to Program

1. The first thing we need to do is open the Terminal application. To do this go to *Applications » Utilities » Terminal.* The screen should

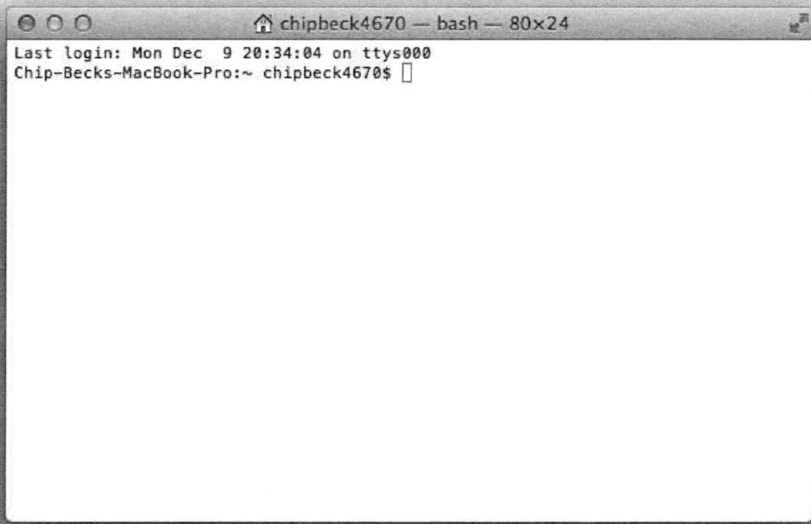

Figure 1.1 - An empty terminal window

appear with the name of your computer followed by your user name.

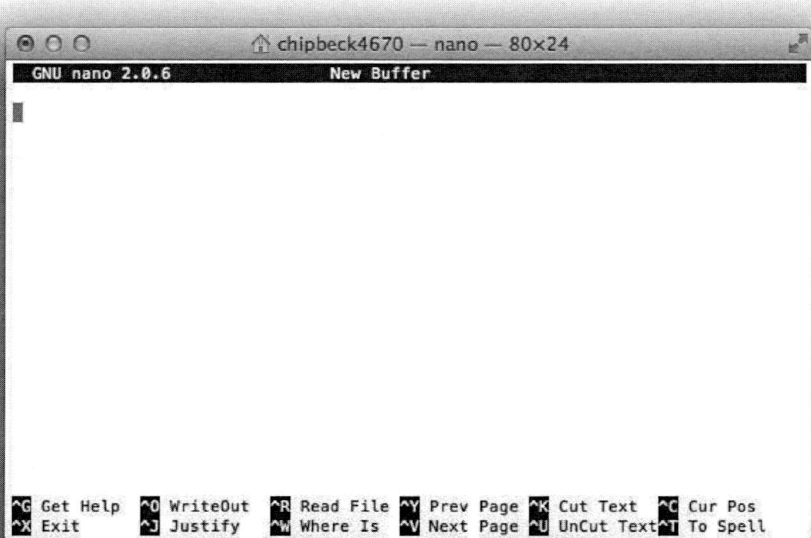

Figure 1.2 - An empty nano editor window

2. Since we need somewhere to write our code, we will use the Terminal's built in text editor called GNU nano. To access this, we need to type:

```
sudo nano
```

and press *Enter*. You will be asked to enter your password (user login password), do so and you will appear at the nano text editor screen shown in Figure 1.2.

Coding Our Program

3. Our Hello World program will be programmed in Ruby, which is a very basic language. Since it is very basic, there is no need for semicolons or curly brackets. What we will output will be a declared string and a declared int. Please type the following into your nano text editor screen:

```
myString = String.new
x = 3
myString = "Hello World!"
puts myString
puts x
```

Figure 1.3 - A nano editor with Hello World code

4. Save your file by pressing *ctrl-x* and then when it asks if you want to save, press *Y*, then it will ask you what to save the file as. Name the file *helloworld.rb* and press *Enter* to finalize and save. This should have taken you back to your main Terminal console.

5. In order for the newly created file to be trusted by the computer, you must type:

```
sudo chmod 777 ./helloworld.rb
```

and press *Enter*

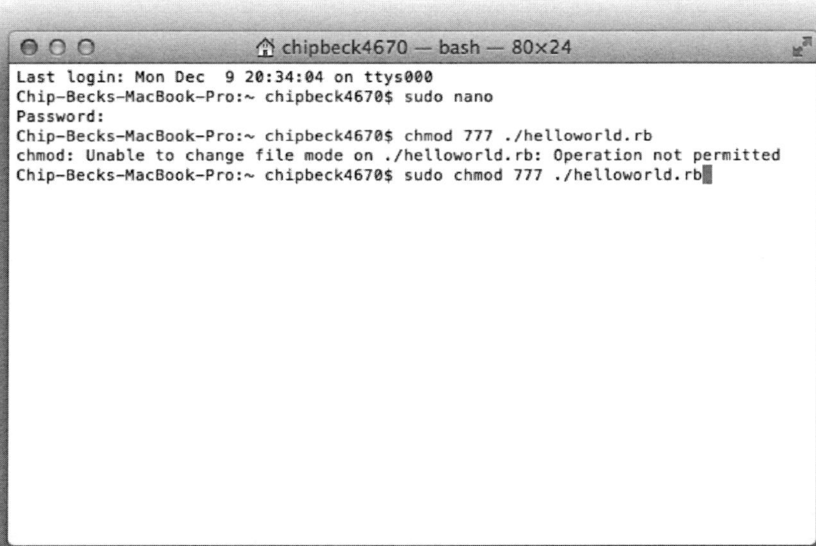

Figure 1.4 - Terminal running chmod command

6. From this point, our new file is now allowed to be accessed by us. We are now ready to run our program via the Terminal. To run the program, we must tell the Ruby executer to run the file. This is done by typing:

```
Ruby ./helloworld.rb
```

and pressing *Enter*. When you have executed the program, in a few seconds the console will display:

```
Hello World!
3
```

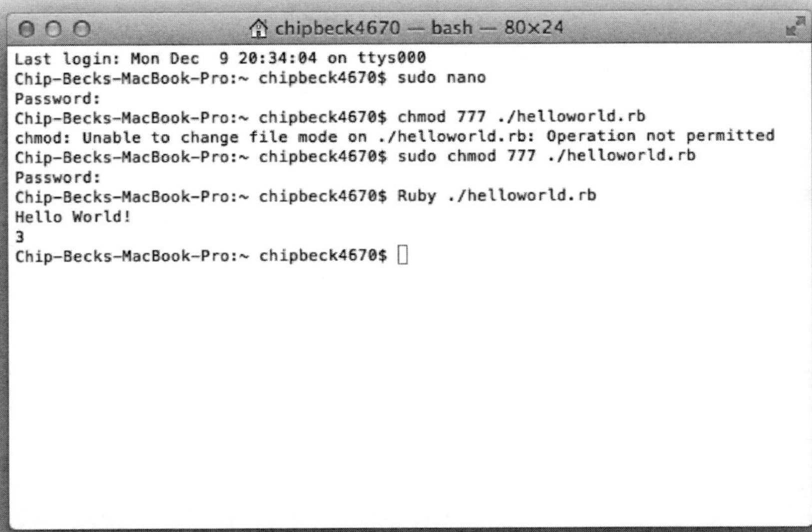

Figure 1.5 - The result of our Hello World program

That's all folks! Your first Hello World program has been completed! It's very basic, but it gets the job done. In comparison with the real world, this is like speaking your first words, or reciting the first sentence of your foreign language class. If you didn't get the results displayed above or want to add more to the program, simply type:

```
sudo nano ./helloworld.rb
```

press *Enter*, and you are back at the editor screen with your old code.

Analyzing the Hello World Program

Understanding what we just typed, and why we typed it is the most important point in the entire chapter. If I don't explain what is going on, it will take a dang long time to grasp the concept of programming. One of the main reasons I wrote this book is because so many "Beginner" books assumed that you know exactly what they did and why, which led to no progress. I want my readers to understand what I did and why. Let's get into it with the first line!

```
myString = String.new
```

This is Ruby's way of declaring a string. In the case of Ruby, you name the variable first, then declare that it is a string and it is a new declaration.

```
x = 3
```

In this line, we are declaring the variable (assumed to be an integer) named *x*. Something very different about Ruby is that it is dynamic, you do not need to declare what type of numerical variable you are using. Whenever you assign something as a number, it automatically assigns the variable as the most reasonable option. In short, this line declares *x* as "3".

```
myString = "Hello World!"
```

This line is assigning the string, *myString*, to equal the phrase "Hello World!", as that's what we want it to say (or else you will be banned from the Hello World Club).

```
puts myString
puts x
```

Since the last two lines do practically the same thing, I'll just do one explanation for both. In Ruby, as with some other languages, *puts* is the command for output. If you were to read this in literal terms, it would say "put out *myString*" and "put out *x*" (not put out as in a relationship, but more of "output"). *Puts* tells the program to display the value of *myString* which is "Hello World!", and on the next line to output the value of *x*, which is "3". What's cool about this is the fact that *puts* can convert *x*, a number, to a string without the need to program anything! All in all, this would save lots of time in converting your numerical variables back and forth from string to number if you are coding a program that uses lots of different variables.

CH. 1 CONCLUSION

Wow! That was a nice size of information that was just tossed out at you as fast as a fiber optic internet download! On completion of this chapter, you now know the basics of general computer programming. You are also now part of my imaginary Hello World Club and consortium of programmers! If you feel you are ready, you can move onto Chapter Two and start learning about the reason you got this book, iOS app development. Please use my checklist to determine if you are ready to move on. If you don't know about an item on the list, that's okay, just look back at what you don't know and try again! These checklists will be at the end of every chapter that covers new topics to make sure you took in what I was putting out.

- You know what a string is (not the cheesy kind) ☐
- You know the difference between an int and float ☐
- You own a Mac (requirement for iOS app development) ☐
- You know what chapter this is ☐

Chapter 2

Setting up Shop

Getting Your Computer Ready to Make Apps

In my experiences, one of the hardest parts of developing apps was getting set up. There are quite a few different files and certificates that you need to install along with reregistering with Apple. This is what many books had left out. If you choose, you can register as an iOS app developer, which grants you the ability to test apps on a device and submit them to the App Store.

2.1 Installing Xcode

First things first, we need to download the software that allows us to make these magical apps. The program that we will be using is called Xcode, an application developed by Apple that allows you to program OS X and iOS apps. Xcode is free to download, so all we need to do is go onto the Mac App Store and search "Xcode". It should be the first item displayed in the results. Click on it and press the button that reads "Free". If you do not yet have an Apple ID, you will need one to download Xcode, so take time to do that now. The download is quite a large one, so please be patient and pop some popcorn, use the restroom, and end world hunger while you wait.

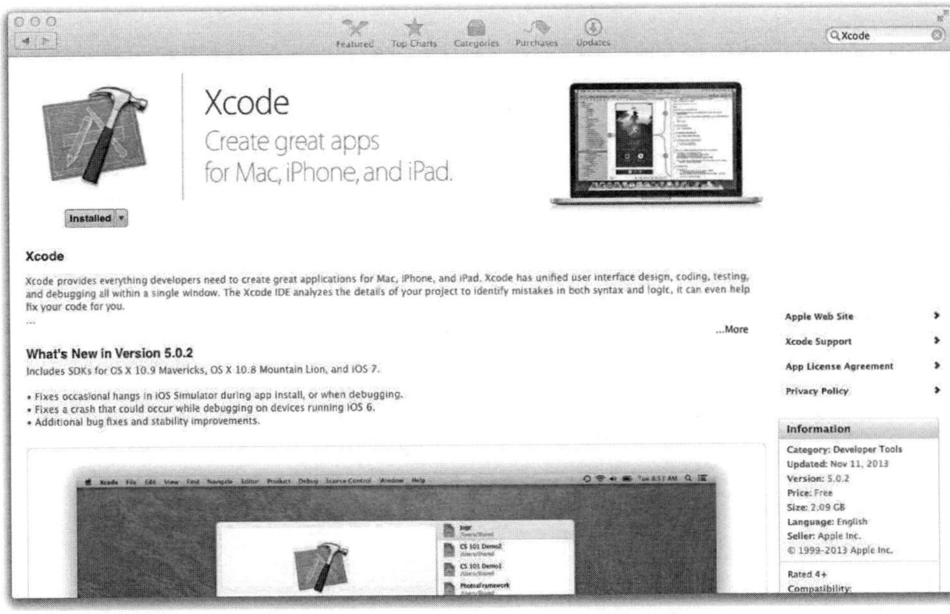

Figure 2.1 - Xcode on the Mac App Store

You're done downloading Xcode, great! Did you get a chance to look into that world hunger thing? Anyways, from here if you are planning on purchasing an iOS App Development License, continue on with this chapter. An individual license is $99 per year. If you are not interested or already have a license installed you can skip onto Chapter Three (you can purchase at any time)!

2.2 Purchasing a Developer License

If you are reading this, you probably made the decision to buy an iOS App Development License. With this license, you will be able to download and install the latest betas, test your apps on your own devices, and submit apps to the App Store. It's a pretty good deal!

The first step in claiming your own developer license is visiting:

https://developer.apple.com/programs/ios/

Then click on "Enroll Now". You will then be taken to a screen which summarizes the process of enrollment. In short, we will be enrolling in the "Individual" program. Please note that to enroll, you are required to provide your legal name, address, and a valid credit card (for purchase of program). Another thing to keep in mind for later is that if you plan on making revenue through either iAds or by

Figure 2.2 - The iOS Developer Program webpage

making your app cost money, you will be required to have a valid bank account, and if you live in the U.S., you will need to provide your Social Security Number (SSN). For those living in a county outside of the U.S., you will need to provide the national identification number that your county provides you with when born. If you don't know your SSN or ID number, ask your parent(s) or guardian(s). Regardless of if you know your SSN or ID number or not, make sure your parent(s) or guardian(s) know that you are going to use it, as that is an extremely important number that you don't want to just give away! Once you have read over the information on that page, click "Continue".

From here, you will need to sign into the developer website with your Apple ID. I would recommend using the Apple ID that you use most often. If you want to have a separate Apple ID for business use, that is fine, although you will have to re-input stored financial details if you had entered them, and your current iCloud account will not be synced with your developer iCloud account. When you determine what you will do, either click “Sign In” or “Create Apple ID”.

The next page will ask you if you are “enrolling as an individual or organization”. Please review the information under the “Individual” column, then at the bottom of the page, click “Individual”. Here on out, follow the form that Apple provides. You will be linked to the Apple Store where you will purchase the developer license. Within twenty-four hours upon purchase, you will be contacted by Apple with an activation code that will contain a link to setup your new account! Once that is complete, you will soon receive an email stating that you now have access to iTunes Connect, the hub for managing apps, iAd, users, contracts, and viewing sales and financial reports. iTunes Connect to be found at:

`https://itunesconnect.apple.com/`

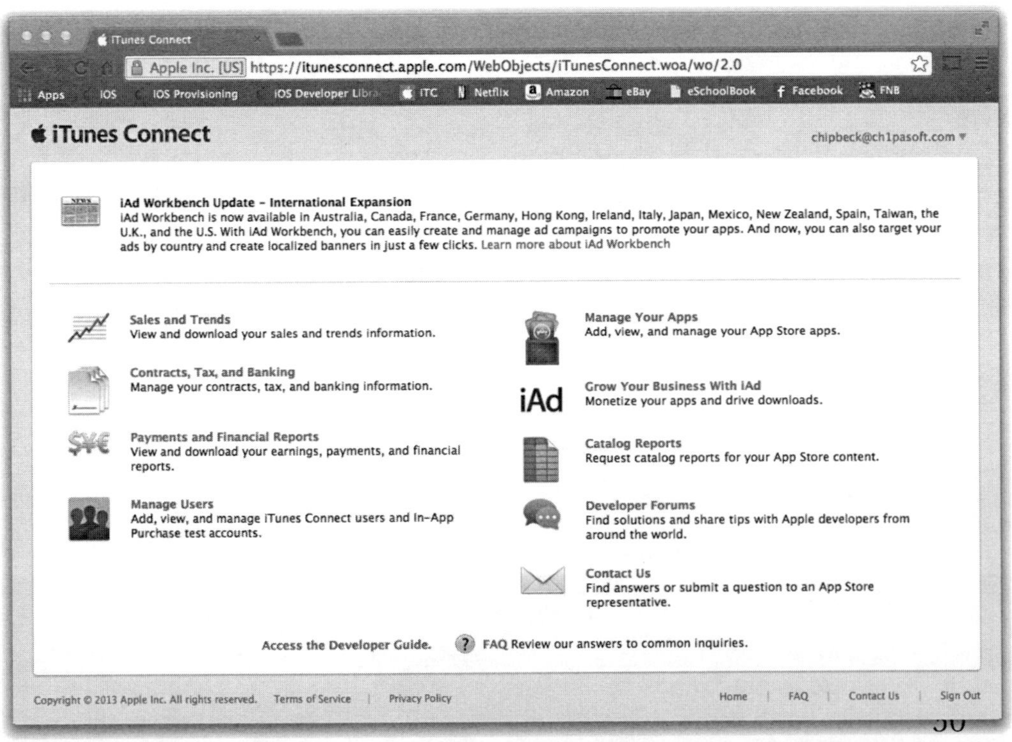

Figure 2.3 - The iTunes Connect website

2.3 Setting up Xcode With Your License

Now that you have your very own, sacred, iOS App Development License, Xcode needs to become tied to your license so that you can get the full functionality of Xcode and the apps that come with it.

The first thing we need to do is open Xcode from the *Applications* folder. Upon opening Xcode, you will be presented with a screen that reads "Welcome to Xcode", followed by recent projects (right) which should be empty for you since this is your first time opening it.

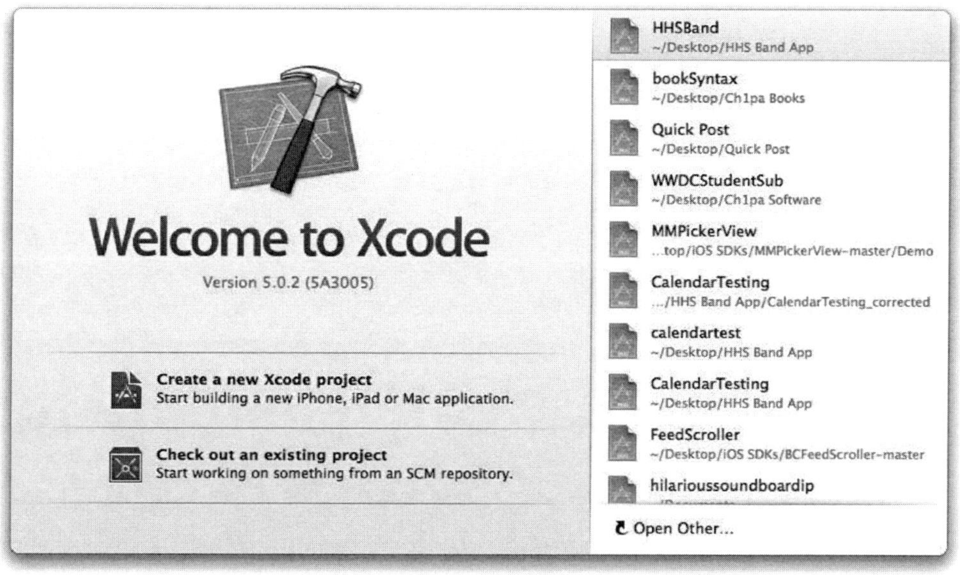

Figure 2.4 - The Xcode welcome window

In order to bind your license and Xcode together, we need to go to the Xcode preferences *Xcode » Preferences*. From the preferences, select the "Accounts" tab. In the accounts tab, you can enter the Apple ID associated with your license. To do this, click the + in the bottom left corner and when presented with options, click "Add Apple ID" (Shown in next figure).

Xcode will then present you with a drop down to enter the Apple ID associated with your license. When finished typing, press "Add" and your account information will then be verified, and tied to Xcode.

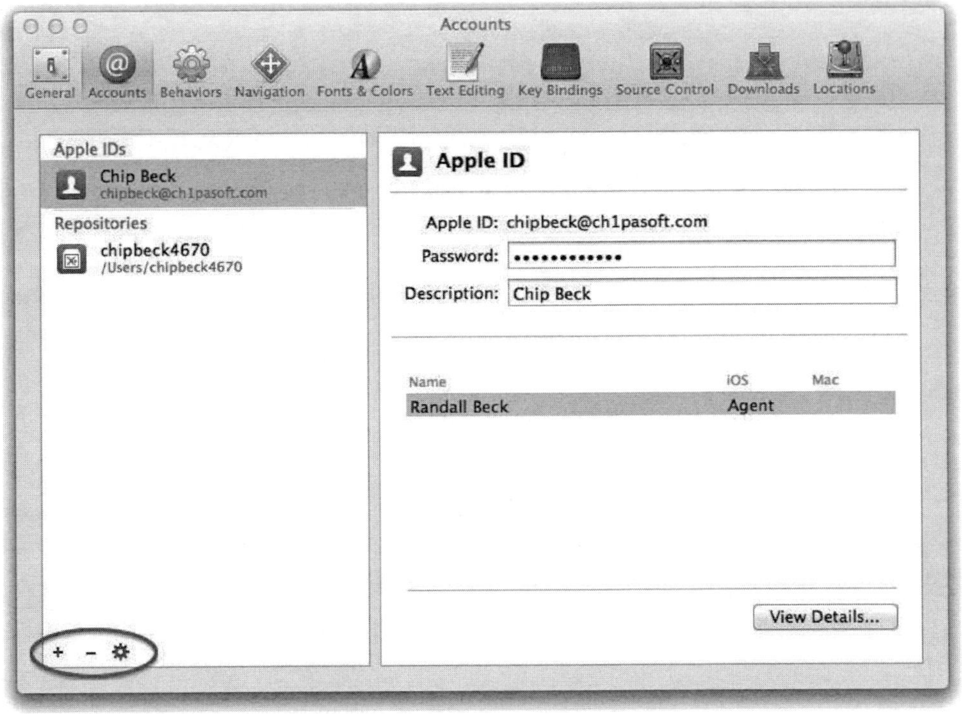

Figure 2.5 - Xcode's Accounts tab

Now that your license has been tied to your account, all testing and building restrictions have been lifted!

2.4 Creating and Installing the Development Certificate

Although all development and testing restrictions have been lifted, to test apps on physical devices, Apple adds an extra layer of security before allowing an app to be transferred onto a device. These security files are called provisioning profiles. Provisioning profiles are sort of like friends, they are a pain, but they tell the jock (Apple) that you are cool and should be let into the group (your device). Provisioning profiles sound cool, but where does one come across a provisioning profile? That's a good question, and we'll come back to that after I explain how to add your devices to the official testing devices list.

Adding a Device to Your Developer Account for Testing

The ability to install your own apps onto your device for testing requires you to add that device's Unique Device Identifier (UDID) onto your developer account. Xcode has the ability to do this built in, so we will use Xcode since it is the easiest approach.

If Xcode is not already open, relaunch it, and at the top menu bar, go to *Window » Organizer* and click on the tab named "Devices". If you have not already done so, plug your iDevice into your computer and wait for device to appear in the column on the left side (under "My Mac"). Once it has appeared, it should display your

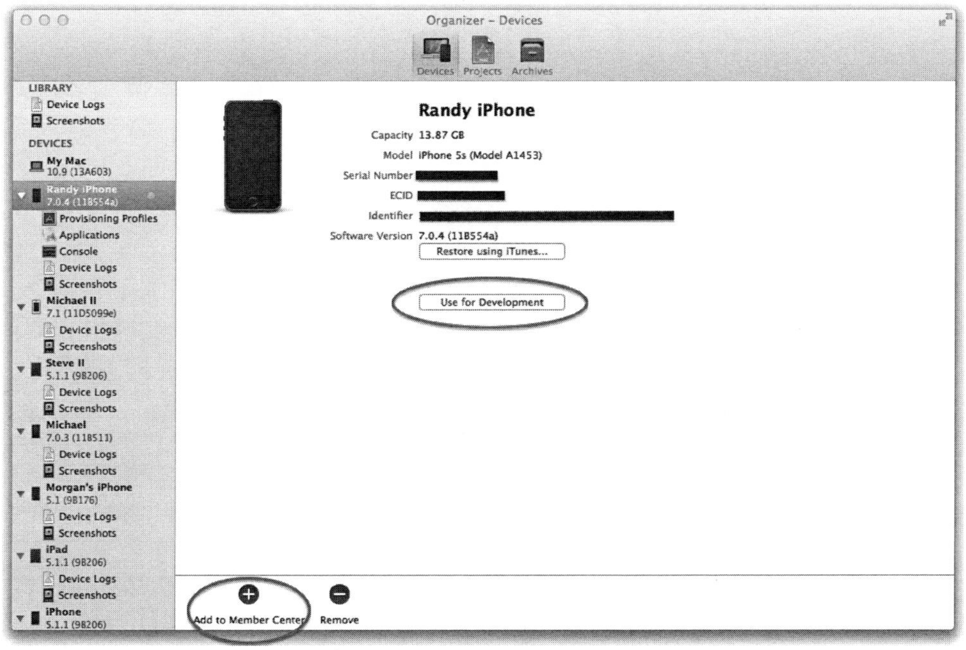

Figure 2.6 - Xcode's Devices tab

device's name, capacity, model, serial number, and a whole bunch of ID's. For your device to accept apps that you build, you must first click on the button that says "Use for Development". This will unlock the developer features on your device. You'll know when your device is finished unlocking whenever a green light appears next to your device's name in the left column. Once you have a green light, click on the + at the bottom of the window that says "Add to Member Center". You may be required to log in with your Apple ID again, do so and your device is ready to test! Repeat this process with any other iDevices that you may want to test your apps on in the future.

Creating and Downloading a Development Certificate

Now that your device(s) can install apps built in Xcode, we can now proceed to create a development provisioning profile. The first step is by going to the Apple Developer Member Center, the place to find all the developer resources you could imagine:

`https://developer.apple.com/membercenter/`

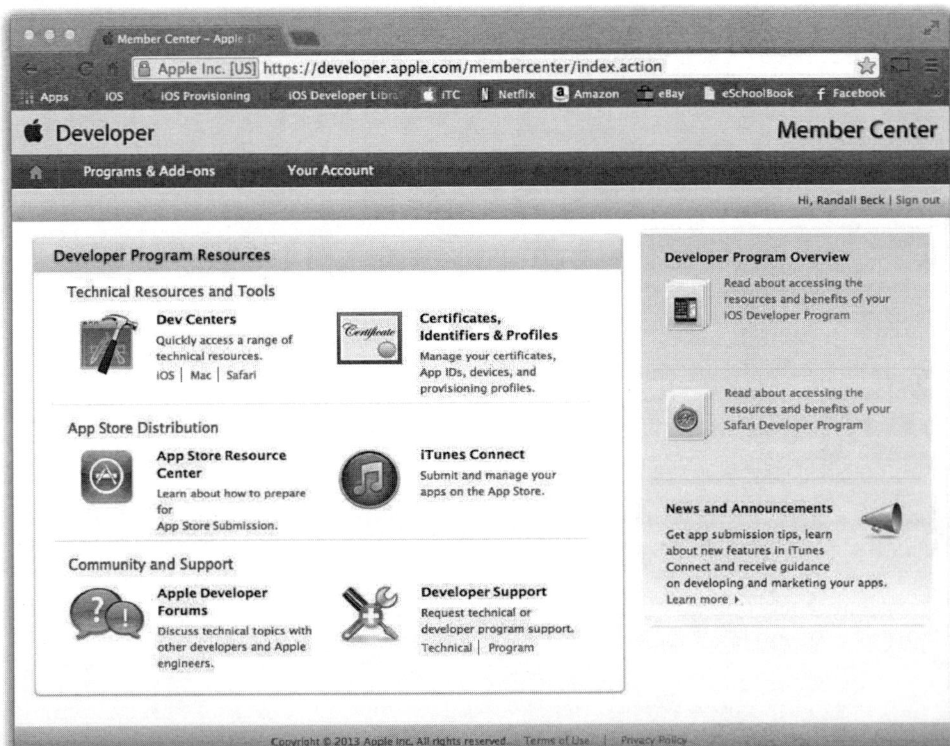

Figure 2.7 - Apple Developer Center website

Log in, and under "Technical Resources and Tools", click on the link "Certificates, Identifiers & Profiles". This should bring you to a page that's right column shows four different folder looking links. The first thing we need to do is make a development certificate. In order to start, click on "Certificates".

From the certificate page, click the + button in the top right area of the window. This will present a form that first asks "What type of certificate do you need?". Since we only want a development certificate at the moment, under "Development" click the circle for "iOS App Development". Proceed to the bottom of the webpage to click "Continue".

The next page will tell you how to create a Certificate Signing Request (CSR) which is a file that will allow you create certificates, which in turn, allow you to generate provisioning profiles. A long and complicated process, I know, that's why I felt the need to write this.

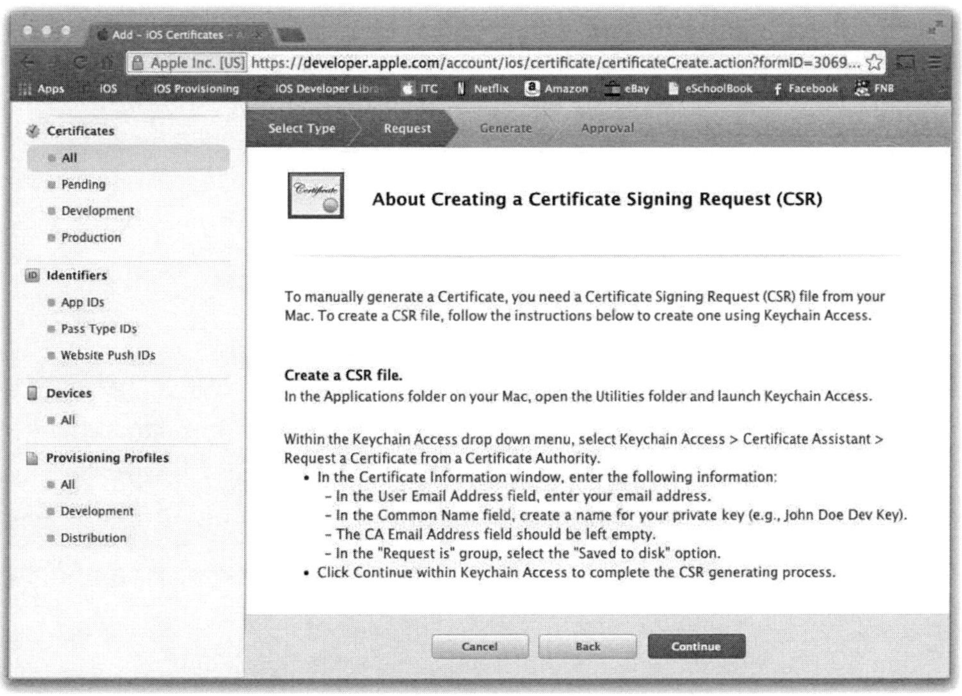

Figure 2.8 - Information on creating a CSR

In short, we will go through the steps to generate a CSR. To start, open the application named Keychain Access. Keychain Access can be found at *Applications » Utilities » Keychain Access.* Keychain Access is

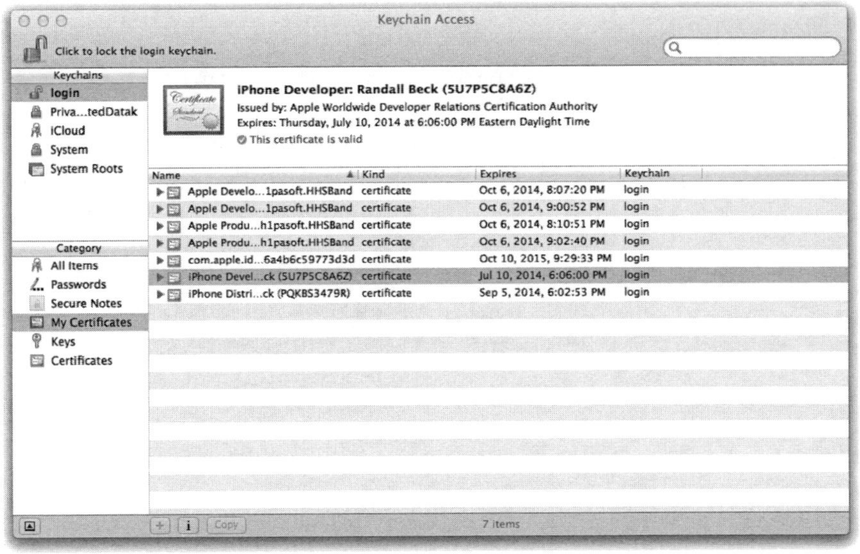

Figure 2.9 - Keychain Access window

an app that manages computer permissions, user passwords, certificates, and computer keys (keys let your computer automatically access "things" that require authentication). Once you have opened Keychain Access, from the top menu bar, click *Keychain Access » Certificate Assistant » Request a Certificate From a Certificate Authority.*

By requesting a certificate from a certificate authority, the Certificate Assistant app should open and appear with fields for your email, name, and the CA's email.

For the "User Email Address", type the email address that is associated with your Apple ID that is tied to your developer license. For "Common Name", to avoid any possible issues, use the legal name associated with your license. Leave "CA Email Address" blank as we don't "actually" have a certificate authority. Click the "Saved

Figure 2.10 - Certificate Assistant information screen

to disk" option as your method of storing the certificate and click "Continue". Whenever you continue, a drop down menu will appear asking where you would like to save the certificate. You can save it wherever you want, but make sure you save it somewhere that you will remember. Once you have successfully saved the certificate, go back to your web browser and click "Continue" to go to the next step in creating a development certificate.

The next step is to upload our newly made certificate to Apple's website. By doing this, Apple's servers will generate a development certificate for us. Go ahead and click "Choose File" under "Upload CSR file" so that we can get this done and start making apps!

Figure 2.11 - Certificate generation page

Once you have uploaded your CSR to the website, click continue to be taken to a page in which you can finally download your development certificate! Good news is that you don't have to repeat this process until next year! Bad news is that you'll have to go through this process every year (you renew your license)!

From the "Certificates" column to the right of the window, click the "Development" row. If you see the newly created certificate, then everything went great. If not, wait a few seconds then refresh as your certificate may still be pending. If it still doesn't appear, then repeat the process above and try again (sorry!).

Once downloaded, click on the development certificate and it should now be in the "My Certificates" section of the Keychain Access program.

If all went well, your new development certificate will give the computer access to building/testing user built applications on devices externally plugged into the computer.

The last step before we can jump into developing some apps of our own is creating a provisioning profile! The provisioning profile lets Xcode know that your app is good (valid) to run on a device and that the user is "fo shizle".

Creating and Downloading a Development Provisioning Profile

To get started, head over to the "Certificates, Identifiers & Profiles" page of the Apple Developer Member Center.

Once you've reached the profiles page, click the "Development" row under the "Provisioning Profiles" column. From there, click the + button in the top right corner of the window. On the next page,

Figure 2.12 - iOS Provisioning Profiles's webpage

click the circle for "Development" and click "Continue" at the bottom of the page. The next page asks you to "Select an App ID". Since we have yet to create any apps, select the "Xcode Wildcard" option. An Xcode Wildcard is an ID that works with any app built in Xcode. If, for some reason, you don't have any App IDs, refer to Chapter Nine Section Three. To save time, throughout this book we will use the Xcode Wildcard to build our apps. Once ready, click "Continue". The next page asks you what development certificate you would like to use. Since we just made a development certificate, we are okay! Select the certificate and press "Continue". The second to last step asks us what devices the profile will work with. I would recommend always using the "Select All" square when making a profile so you are sure you don't leave anything out. Once you have selected your devices, click "Continue".

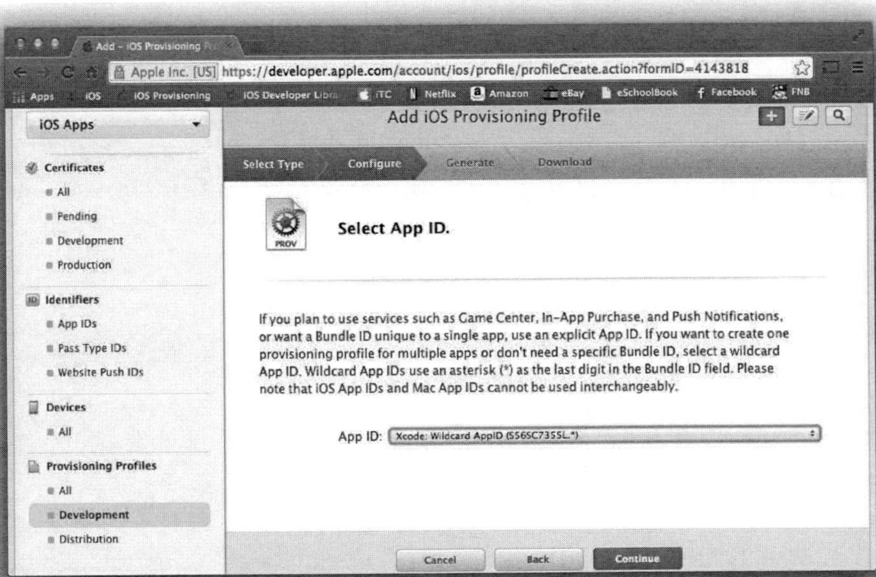

Figure 2.13 - Requesting a provisioning profile

Ok, last step of this section and chapter! All we must do now is simply name our profile. You can choose whatever name you like, although I would recommend choosing something memorable for the future. When you type a name, you can finally click "Generate".

Wait a few seconds (depending on your internet connection) and you will be taken to the download page. Click "Download" and once downloaded, click on the file, wherever it is located, to add it your computer's list of valid provisioning profiles.

CH. 2 CONCLUSION

It looks like we are FINALLY ready to make apps! The next chapter goes over the basics of Objective-C, the language of iOS apps, and you get to make your first app! Please make sure you understand the concepts and/or have done the items in the checklist before moving on.

- You have an iOS App Development License ☐
- You have an iOS development certificate ☐
- You know how to make a provisioning profile ☐

Chapter 3

Introduction to Objective-C and Xcode

You've made it! You've made it to the blessed chapter where you get to learn all you need to make a Hello World app for iOS! I know this doesn't sound like much, but with this basic knowledge, you will get a small taste and grasp of what Objective-C and Xcode are all about. In this chapter you will also learn about the structure of Objective-C, methods of declaration, voids, outlets, and actions. Sound confusing? That's great and also why you're here! Let's jump into iOS app development!

Figure 3.1 - The Xcode welcome window

3.1 Creating an Xcode Project

Xcode Integrated Development Environment (IDE) is where all of the magic happens. Xcode's IDE is best suited for beginners as it features the ability to drag and drop functional design assets, and even drag and drop some skeleton/template code.

To get started, open Xcode from the *Applications* folder. This will pop up Xcode's clean welcoming window. On this window, click "Create a new Xcode project" to be presented with basic template options.

Xcode comes standard with a nice variety of templates that provide the basic structure and foundation of your app. Let's go through and see what each template has to offer.

Comparing Xcode Application Templates

	What's Included	Difficulty
Single View	View, View Controller, Storyboard	Easy
Utility	Main View, Alternate View, Info button (to switch views)	Easy
Tabbed	Tab Bar, Tab Bar Controller, View Controllers, Storyboard	Medium
Page-Based	View, Page View Controller, Storyboard	Medium
Empty	App Delegate, Empty Window	Medium

	What's Included	Difficulty
Master-Detail	Navigation Controller, Split View, Storyboard	Medium-Difficult
SpriteKit	SpriteKit Starting View	Difficult
OpenGL	OpenGL ES View, Timer	Difficult

Project Options

Since we will use this project for our Hello World, lets select the "Single View" template option. This template is a very basic, barebones template that provides us with just what we need to accomplish our task.

The template options page is where you name your project, give it a company ID, and let the project know who is creating it. For "Product Name", simply enter in your project's name (spaces

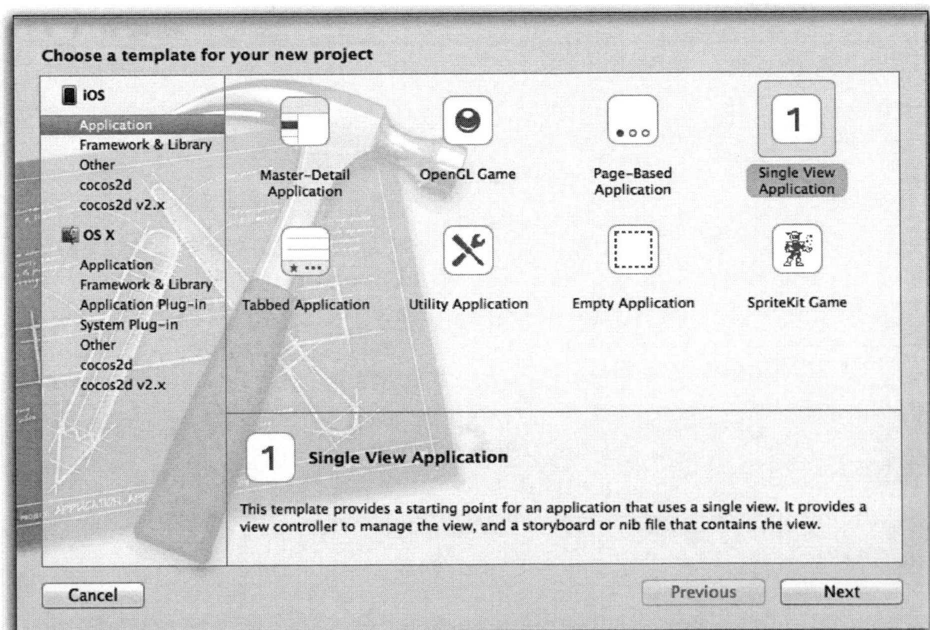

Figure 3.2 - Xcode's various templates

allowed). The "Organization Name" field is the field which has to do with your project's copyright. According to U.S. law, a copyright is automatically created at the moment that a unique piece of digital content is created, whether an app or a song. For "Organization Name", either place your full legal name or, as in my case, your company's name. The "Company Identifier" field is a very important field if you are planning on submitting the app to the App Store. This is due to the fact that no two apps on the App Store can share the same bundle identifier, the app's user created ID. For your identifier, I'd recommend using your full name, or a combination of numbers and letters generated in a password generator in order to prevent the remote possibility that someone else is using your company identifier. In theory, all of your apps should share the same company name, although, it is not required. The reasoning behind bundle identifiers is that it represents the creator(s)'s website URL, but backwards. For example, apple.com's bundle identifier for a project titled "Pages" would be com.apple.Pages. As part of the company identifier, you should have a domain such as "com" or "net" preceding the actual name. Below the "Company Identifier", should be your project's bundle identifier. Again, I cannot stress

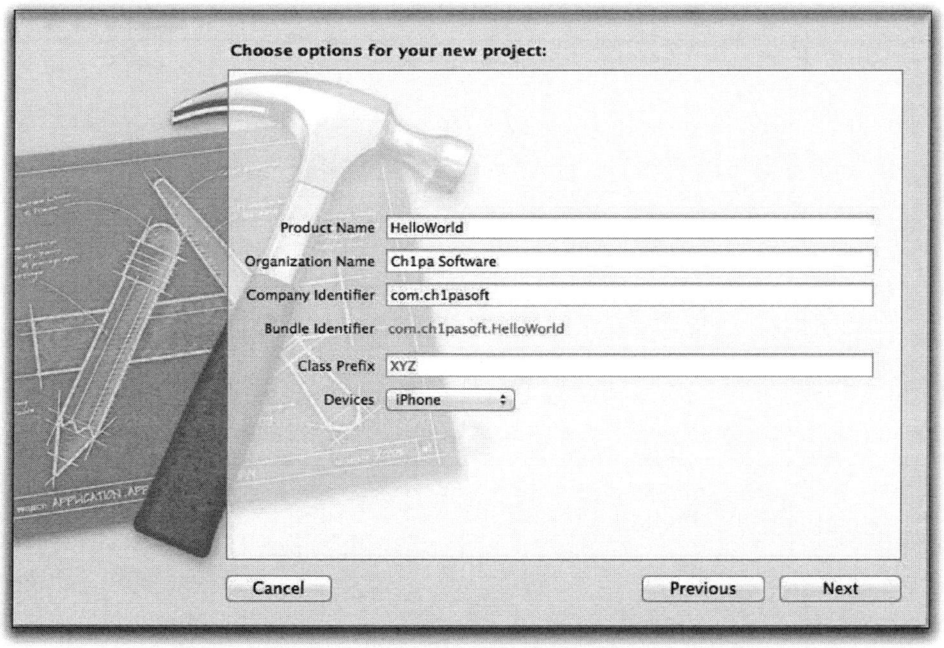

Figure 3.3 - An Xcode project's options page

enough how important your bundle identifier is whenever it comes to app submission, as there is no way to change it after the project has been created. To show how much I stress this, I'll tell you my story.

Why You Need to Have a Unique Bundle Identifier

It was finally done, my first app was ready to be submitted to the Apple for approval. I went to upload my app through Apple's Application Loader, and pressed "Submit". When I submitted it, I received an error message back instead of a success confirmation. I was shocked and read the error details stating that someone else already had an app on the App Store with the same bundle identifier as I did! Upon reading this I realized that I would have to completely start and program the exact same app again as a separate project due to my bundle identifier! To make things worse, because my app information on iTunes Connect was tied to my bundle identifier, I would have to change the name of the app and re-enter all of the legal information as a new app! I had just wasted most of my summer creating an app that wouldn't even get past the submission authenticator. This is why you MUST have a unique bundle identifier.

Project Options (Cont.)

The "Class Prefix" field is not required to be filled out. What this does is simply place what you put in this field as a prefix to every class you create. The main reason you would use this is in the case that you are creating your own framework, SDK, or custom control that you plan on making open source and want to save time, although, you may use this for any project if you please. For this project, we will leave the field empty. The last option is what device(s) you want to support. The options are iPhone, iPad, and "Universal".

The iPhone option is optimized for iPhone and iPod Touch but also supports the iPad. The iPad option only supports the various iPads, and the "Universal" option supports, and is optimized for the iPhone, iPod Touch, and various iPads. Once selected, this is the only option that still can be changed once the project is created. For our Hello World, we will select iPhone option. Once you're sure that all of your options are correct, click "Next" to continue.

Now that all of our options are in place, we need to create the project. On the next screen you will be asked your save location. I'd recommend creating a dedicated folder in either your *Documents* folder or *Desktop*. Once you click create, a folder (not the one you might have just made) will be made containing a .xcodeproj file and the project's supporting files. The project should now open. Leave it open as we will be using it throughout the chapter.

3.2 Objective-C Structure

It is now time to examine the structure in which code in Objective-C is written. We will also be coding our Hello World project throughout this section as you learn about the structure of Objective-C code. Objective-C is an OOP language which means that the developer initializes classes along with variables and strings that interact together in some way or another.

Header Files

In Objective-C and many other programming languages, header files (.h) are files that contain pre-named variables, functions, actions, etc. that can be used in method (.m) files. The purpose of header files are to both predefine some items, and/or pre-name items that will be used throughout the project. In your Hello World project, click the file named "ViewController.h" so that this appears:

```
#import <UIKit/UIKit.h>

@interface ViewController : UIViewController

@end
```

Every *ViewController* file created will, by default, look like the code above. *<UIKit/UIKit.h>* is imported automatically by Xcode, as it is the main library of code for the iOS SDK, and the only library we need to code a Hello World in Objective. The following *@interface ViewController: UIViewController* code is the code that defines what kind of class the file needs to represent, a UIViewController in this case. As with most, if not all OOP languages, *@end* is used to end the code in a file.

It is possible to split a header file into two parts, one part for defining classes, variables, and outlets, and another part for defining actions and functions. For most projects, including this one, we will want to do this. In order to split it, place an open curly bracket ({) after the line *@interface ViewController : UIViewController*, and a closed curly bracket (}) the line before *@end* so that it looks like this:

```
#import <UIKit/UIKit.h>

@interface ViewController : UIViewController
{

}
@end
```

Using this method, the section for classes, variables, and outlets is in-between the curly brackets, and the section for actions and functions is after the closed curly bracket and before *@end*. The text in the Hello World project will be represented through an IBOutlet UILabel and a string. The IBOutlet is added before UILabel because this UILabel is going to be seen by the user as part of the interface and is built through Interface Builder, hence the “IB” prefix. This is just a sampler on outlets and classes, so don't worry, we'll learn more about this later. For now, make your code look as such:

```
#import <UIKit/UIKit.h>

@interface ViewController : UIViewController
{

    IBOutlet UILabel *myLabel;
    NSString *myString;

}
@end
```

You may notice that a little circle appeared to the left of line counter next to the line containing *IBOutlet*. This always occurs on lines that have to do with Interface Builder. Whenever you use this object in Interface Builder, the empty circle is filled and darkened.

Method Files

Method (.m) files are where all the meat of the code is (some other people/resources refer to these files as message files instead of method files). This is where you get to really tell the program what to do, and what everything you initialized in the header is supposed to do. Method files are partly dependent on what frameworks and other classes you initialized in the corresponding header file. To open the default method file, click on the item named *ViewController.m*. The default file should look like this:

```
#import "ViewController.h"

@interface ViewController ()

@end
```

```
@implementation ViewController

- (void)viewDidLoad
{
    [super viewDidLoad];
     // Do any additional setup after loading the
view, typically from a nib.
}

- (void)didReceiveMemoryWarning
{
    [super didReceiveMemoryWarning];
    // Dispose of any resources that can be
recreated.
}

@end
```

As you may have noticed, the first line is to import our *ViewController.h*. By doing this, all frameworks, classes, functions, etc. are brought over to the method file so you don't have to re-import or re-initialize anything from the header file.

From *@interface ViewController ()* to the first *@end* can be ignored, although not deleted, as it is still important to the project. Where we want to start is *@implementation ViewController*. What this line does is simply call to implement the file as "ViewController".

-(Void)viewDidLoad is where the fun begins. The function of *viewDidLoad* is the first thing in the file that loads when the app fully launches. In here, you put everything that you want to run before the user visibly sees the app's interface. At the beginning of every default, void function is a line of code, *[super nameOfFunction];*. It is always a good idea to do your coding after this line, as it allows Apple's code setup to load before your code does. We'll come back to this later for the Hello World code setup.

The only other default function that is included initially is the function *-(void)didReceiveMemoryWarning*. The purpose of this is to free up memory that is not currently needed whenever the app senses low memory. In this void function, the *super* command runs Apple's default script to free as much memory as it can, but not in the apps currently running. This leaves you to un-initialize something that is set to be recreated later, or is not needed anymore. The only time you would truly mess with this function is if you are developing a very memory intensive app such as a game, or video streaming app. For our tiny Hello World project, we will leave this function alone.

As with the "end" of every file of code, an *@end* is guaranteed to follow. This is to tell the program that it is finished with this section.

3.3 Coding the Rest of the iOS Hello World project

It's time to finish off the Hello World project. Return to the file *ViewController.h* and check that your code matches the code below:

```
#import <UIKit/UIKit.h>

@interface ViewController : UIViewController
{

    IBOutlet UILabel *myLabel;
    NSString *myString;

}
@end
```

If your code doesn't match, copy this code down, and replace your old code. If it does match or if you correct your code, continue on and go to the *ViewController.m* file.

We want to have the text "Hello World" printed in both the system console and on the screen. In order to do this, we will set the string that we already created in *ViewController.h* equal to "Hello World". To do this, go to the next line below *[super viewDidLoad];* (feel free to delete any comments // that you don't want) and add the code:

```
myString = @"Hello World";
```

Now, the string, *myString*, is equal to the phrase "Hello World"! In many OOP languages, to set something equal to a text string, you need to encase the text with @"". Also notice that in Objective-C and other OOP languages you end each line (not including comments) with a semicolon (;). We should put our string to some use! Since the log (Objective-C term for system console) can output strings, lets make the log output our string. In order to do this, add this code below your string declaration:

```
NSLog(myString, nil);
```

Time to explain the logging function! You may be wondering what the "NS" prefix is doing before everything. The simple answer is that it stands for "NeXTSTEP", part of NeXT, the company founded by Steve Jobs, one of the original users of Objective-C who eventually purchased the language. The "NS" prefix is completely unique to Objective-C and Objective-C only. Now back to the code! The function's skeleton is *NSLog(NSString, *format);*. What this means is that NSLog must output a string followed by a comma, then with it's format (such as an error code). In our use, we want to log *myString*. To do this, we simply set the *NSString* to *myString*, and since we don't need any special formatting, we set the **format* section to *nil*, or nothing, empty, or 0 (in programming terms; definition depends on usage).

The last thing we want to do, is make the text "Hello World" appear on the user's screen. To make this happen, we must use our UILabel (the "UI" prefix stands for "user interface") , *myLabel*, to make the text visible. Under your logging line, type:

```
myLabel.text = myString;
```

By doing this, the label's text will be set to the value of *myString*, "Hello World"! For many "UI" prefixed items, you can access commands such as *myLabel.text* by following the name with a period and the command. For example, you can set text color, alignment, alpha (visibility), etc. by following the name of the item with a period and a command. Since we're talking about it, let's set the text color to blue by typing:

```
myLabel.textColor = [UIColor blueColor];
```

UIColor

(*Optional) Since the property of *textColor* is UIColor, you must set the color to be a UIColor, and in our case, blue, *blueColor*. Statements like *[UIColor blueColor];* must be surrounded with brackets ([ ]) because you are defining what value you want to set to the property to, in this case, UIColor. There are infinite possibilities with color. Many colors are built in, and you just need to type the name followed by "Color" (ex. *purpleColor, redColor,* etc.). Other colors must be user defined in RGB values. This is showed off by editing the value past UIColor:

```
[UIColor colorWithRed:0.6 green:0.7 blue:0.3
alpha:1.0];
```

The max amount each color can be defined as is *1.0*. Since you probably don't have RGB values memorized, you can easily look online to find a color. The only problem is that Objective-C RGB is maxed at 1.0, while the normal version of RGB is maxed at 255! In order to convert the normal value to the Objective-C value, divide the normal value by 255. For example *255 / 255 = 1.0* (yes, there will probably be some decimals).

3.4 Finishing Hello World With Interface Builder

Now that we have *myLabel* set to *myString*, we need to make it actually appear! The way you build your app's interface in Xcode is through Interface Builder. Way, way back, Xcode and Interface Builder used to be two different and confusing applications, but they are now combined in Xcode! To access it, click on the file named *Main.storyboard.* Apple introduced the "Storyboard" design interface not too long ago, and it is a very simple way to manage and design your app. Once the file has loaded and appeared on your screen, you will be presented with a blank iPhone screen.

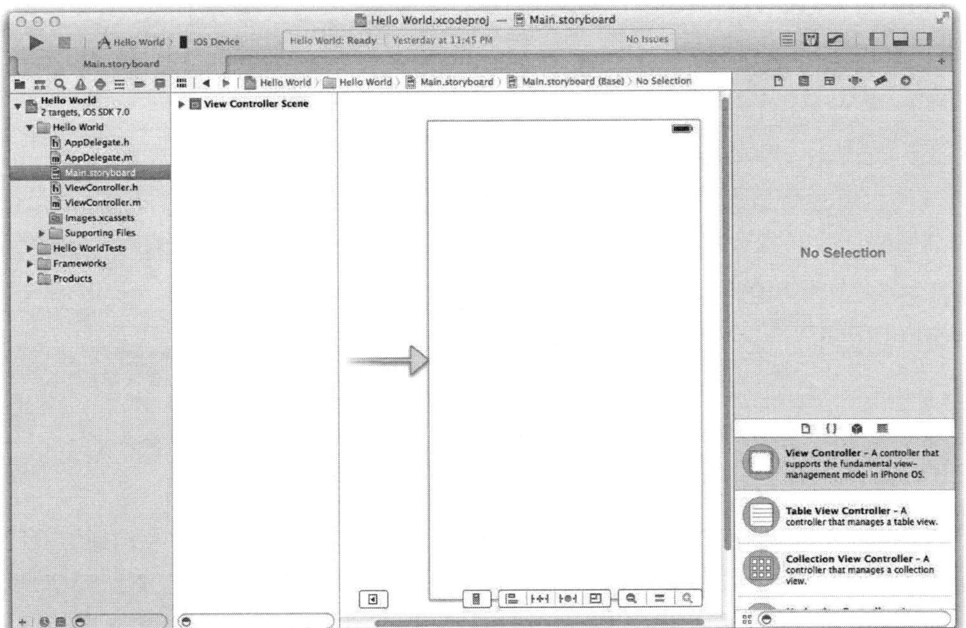

Figure 3.4 - Blank iPhone template

To the side of the window, you will notice options like "View Controller", "Table View Controller", etc.. Scroll down the list until you come upon "Label" (towards the top). Drag and drop "Label" anywhere onto the iPhone template. Once that is completed, click on the new label. Look to the row in the top left that reads "View

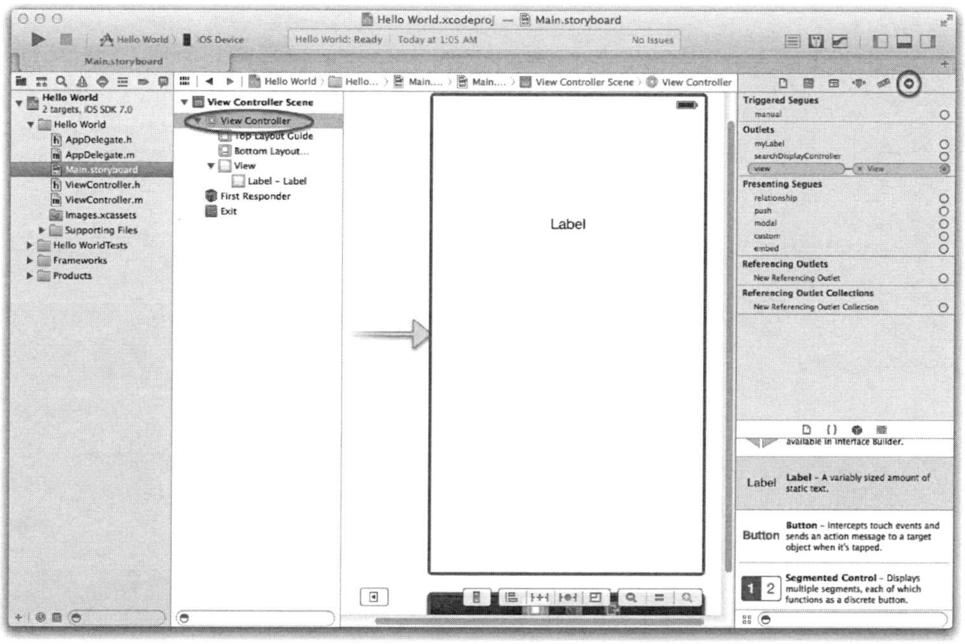

Figure 3.5 - Storyboard with label

Controller Scene", and click on the item that says "View Controller" and has a yellow-orange icon next to it.

From there, click on the right arrow-looking icon near the top right corner of the window. You will be presented with rows for "Triggered Segues", "Outlets", "Presenting Segues", "Referencing Outlets", and "Referencing Outlet Collections". We are looking for the "Outlets" section. Notice that the UILabel that we previously defined, *myLabel*, is located there! Click and drag the dot in the far

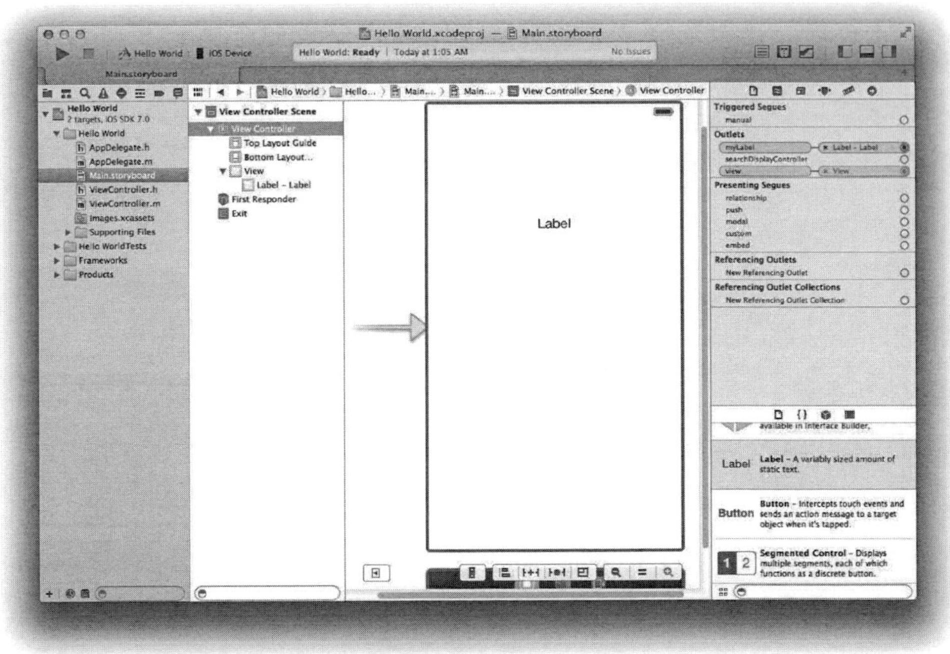

Figure 3.6 - Storyboard with myLabel connected to the label

right to the label, which should then be highlighted, and release the click. If the little dot in the far right is now darkened along with the outlet, *myLabel*, then everything went well, the Hello World project has been completed, and we are ready to compile and run!

3.5 Building and Running Projects

Now that you've completed the project, I'd imagine that you want to see it in action! This section will be divided into two parts, one that shows you how to run the app in the iOS Simulator, and the other shows you how to run the app on a physical device (if you have a developer license).

Building for the iOS Simulator

When installed, Xcode comes with a variety of debugging and testing tools, including the iOS Simulator app. A good device to test on is the option "iPhone Retina (4 - inch)". This option includes devices such as the iPhone 5, iPhone 5C, iPhone 5S, and later on. To select this, near the top left corner of the window, click the title that may say "iOS Device", "iPhone Retina (4 - inch 64 bit), "iPad", etc., and choose the option that says "iPhone Retina (4 - inch)". We are not choosing the 64 bit version for two reasons. One, the 64 bit version is exclusive to the iPhone 5S+, and two, some code and options in 64 bit are different than those in 32 bit.

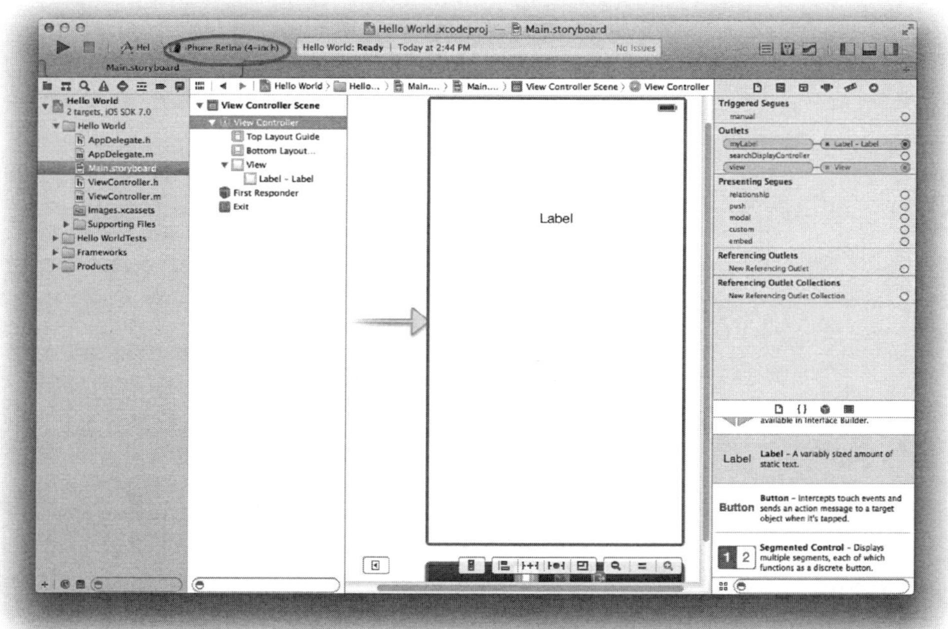

Figure 3.7 - iPhone Retina (4 - inch) build option selected

From here, either press the run (play-looking) button in the top left corner, or hold the keys ⌘*R* , and Xcode will first build the project, and then the iOS Simulator app will start and eventually display (it takes awhile!) the app with the text "Hello World" where you placed your label, and in the color that you chose! The text will probably be cut off at "He…". If this happens, it's okay, just go back to Xcode, click on your label and drag the corners to make it bigger!

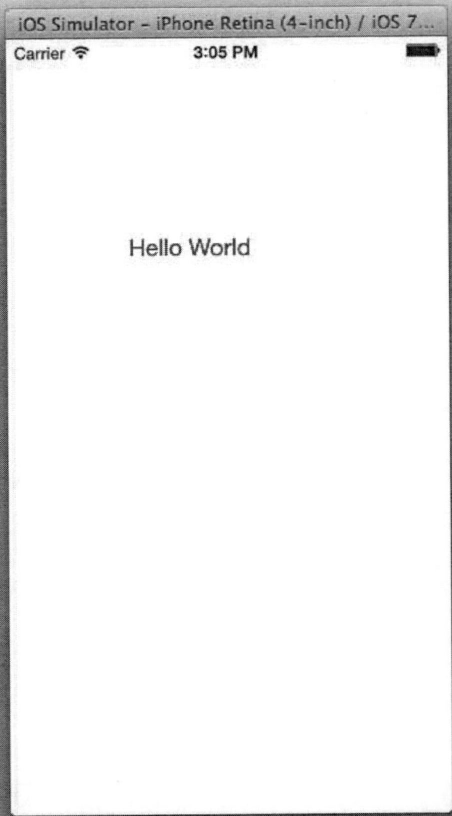

Figure 3.8 - Hello World in the iOS Simulator

You may be wondering where the log is, it's back in Xcode, towards the bottom of the window. If coded correctly, the log will display the text right at the very bottom.

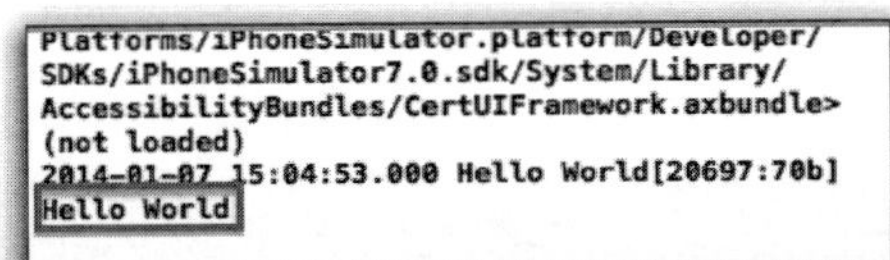

Figure 3.9 - iOS Simulator log

If the app didn't run, make sure that the build scheme is set to "iPhone Retina (4 - inch)" and not "iOS Device" or the name of an iOS device that is connected to your computer.

Congratulations on completing your first iOS app!

Building for an iOS Device

If you have a developer license, let's get this app on your device to show off to your friends that you can make an app!

We first need to change the build scheme in the top left to "iOS Device", and plug your iDevice into your computer. By doing this, "iOS Device" changes into the name of the device you just plugged in (it may take a while to appear). The next step is clicking the project item titled "Hello World" and has a blue project icon next to it. Click on the "Build Settings" bar item and scroll down to the "Code Signing" section. From there you will want to change every

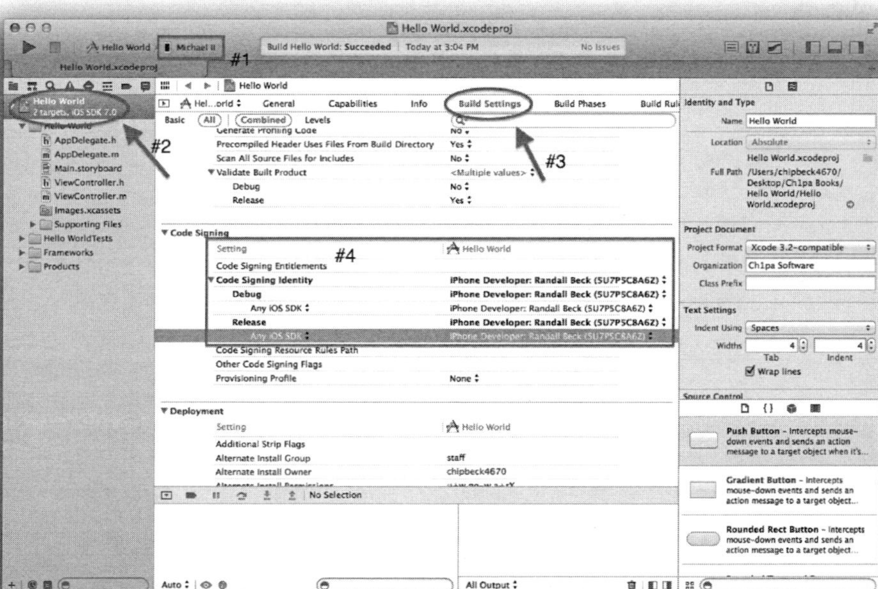

Figure 3.10 - Settings for building on a device

part of the "Code Signing Identities" to the development certificate that we made in Chapter Two.

Once the settings are changed, the app is ready to be built. To build, either click the run (play) button in the top left corner, or hold in the keys ⌘*R*. Once it has started building, Xcode will ask if you will allow it to be built, you must either click "Always Allow" (to be never asked this again), or click "Allow" to only allow this one build.

This is it! Hello World should now be on your iDevice when Xcode is finished building! Congratulations for successfully completing the Hello World project.

Build Troubleshooting

If you can't find your development certificate, re-download it from the Developer Member Center and drop it into Xcode.

If the project doesn't build, make sure all identities are set to the provisioning profile as shown in Figure 3.10.

If the error is device specific, make sure that you have registered the device for development as shown in Chapter Two.

If the project doesn't build due to errors in the code, recheck your code to see if it matches the code above.

If your problem doesn't match any of the above issues, feel free to email me at ChipBeck@Ch1paSoft.com, or go to StackOverflow.com and look up your issue.

CH. 3 CONCLUSION

It looks like we finally got to program an iOS app! I know this is just a basic app that displays text, but it is a giant milestone in submitting an app to the App Store! I hope that everything worked well and that you are understanding programming a little better, but don't worry if you aren't as there is still plenty to come.

- You know what header and method files are ☐
- You can type code that I supply ☐
- You successfully ran the app on your device of choice ☐
- You're ready to learn more methods of content presentation ☐

Chapter 4

Presenting Hello World Differently

Learning New Methods of Presenting Information

Keeping people interested in something is a tough task. In this book, for example, I try to throw in relations to the real world, or make a joke every once in a while to keep you interested. This is the same with app design, although you want consistency, you still need to make the app "different" in some way or another. Instead of presenting everything with a message of text, you can use different types of views, alerts, and buttons to perform actions and display content differently. Although presenting this chapter was not part of my original plan, I feel it is important that you learn about the various methods of presenting content early on so that you can apply them by yourself in later chapters, or even add your own flare to one of the chapters found in the book.

This chapter contains many different, short sections, each containing their own method of presenting the phrase "Hello World". I know… Hello World projects are boring and you want to get into making the next best-selling app, but trust me, this is worth the time. If you learn and understand the concepts in this chapter now, it will make the rest of the chapters seem simple, and you'll think "Why didn't I think of that!?". At some point in your iOS app programming journey you'll say and/or think just that. I encourage you to use your own names, and move around the code a little bit in this chapter so that you aren't intimidated to create your own code later on. By only copying what I'm putting down, you don't get the full effect like you would by doing your own coding and naming.

4.1 Hello Void Functions

You may recall the term "void" (not *Dark Void* or some sort of nightmare). We used the void function *viewDidLoad* to set the value of the string, while also printing that string in the log and changing the label to equal the string's text. Developer made void functions must be called (accessed) via a simple line of code each time you want to use it. It is great to save space and run multiple lines of code at once, and is even better if you use the code a lot throughout the program.

Even though we aren't going to call the void assigned with our code more than once, it's still useful to learn how to make a void function.

We'll get started by reusing the Hello World project we created in the last chapter. In case some sort of tragic event happened to the project (such as deletion), or you're starting to read from here, the code for the header (.h) file is:

```
#import <UIKit/UIKit.h>

@interface ViewController : UIViewController
{

    IBOutlet UILabel *myLabel;
    NSString *myString;

}
@end
```

We're going to want to initialize a void function in the header file so that the method (.m) file understands what we need. In order to do this, go to the line after the closed curly bracket (}) and before the final *@end*. The section after the closed curly bracket is the section where you initialize functions such as the void function. Here, you will want to type:

```
-(void)helloWorld;
```

Sadly, the functions in this section are initialized differently that in the first section. Functions begin with a hyphen (-), are enclosed in parenthesis (( )) and followed by the desired name. This is the only thing we have to do in the header file, so let's go to the method file.

Now we want to set up the function *helloWorld* in the method file so that we can put it to good use. Doing this is very similar to initializing it, the only difference is that you replace the semicolon with curly brackets:

```
-(void)helloWorld {

}
```

That's it to get the function set up! Now lets get some code in there. The code is the exact same as we used in *viewDidLoad,* so you can either highlight and drag the code to the void function, copy and paste it, or just retype the code. The finished function should like like this:

```
-(void)helloWorld {

    myString = @"Hello World";
    NSLog(myString, nil);
    myLabel.text = myString;
    myLabel.textColor = [UIColor blueColor];

}
```

If you still have code in the *viewDidLoad* function, delete it (not *[super viewDidLoad]* though), as the new function should be the one performing the action, but first we need *viewDidLoad* to call upon *helloWorld*. Calling is very simple in Objective-C, the code is always *[functionName self]*. In this case, the code will be:

```
[self helloWorld];
```

That's it! The project is now ready to run with a brand spankin' new function!

The command *self* in this sense is that the program is telling itself to run the *helloWorld* function. For example, if something else was supposed to run the *helloWorld* function, the code would be: *[otherFileName helloWorld];*.

Now it's time to run. Run the project in the same fashion as you did in Chapter Three. If app looks the exact same as it did in Chapter Three, then the program ran correctly. If it didn't, recheck the code to make sure that it's the same as mine.

The *viewDidLoad* and *helloWorld* functions should look as such:

```
- (void)viewDidLoad
{

    [super viewDidLoad];
    [self helloWorld];

}

-(void)helloWorld {

    myString = @"Hello World";
    NSLog(myString, nil);
    myLabel.text = myString;
    myLabel.textColor = [UIColor blueColor];

}
```

4.2 Hello IBAction

The IBAction function is the void function's interactive brother (or sister if you prefer). The method of declaration is practically the same as void, and the method of defining it's actions are also the same. The only big difference is the fact that IBAction *is* an action that the method of accessing the function is by assigning the function to something in Interface Builder. It's time to jump (not literally) on this IBAction thing.

We'll want to reuse the code that we used in the last section, so in the case that the file is gone, or you are just joining now, please go back and copy the code into an Xcode project so we can get started!

First thing's first, lets declare an IBAction named whatever you want (but not the same as any of the other names already used) in the same section where the void function was declared.

```
-(IBAction)hello;
```

That's the simplicity of IBAction. Although some programmers follow the name with an *id:sender* command, it is never necessary unless specifically asked for. It's already time to move to the method file!

Just as you did with declaring the void function, you will do the same, and replace *void* with *IBAction*. The code is:

```
-(IBAction)hello {

}
```

In this section, we'll implement the IBAction feature by making a label disappear whenever we press a button associated with *hello*. To make this happen, we are going to want to put a line of code into the IBAction that makes the label that we declared earlier become hidden. This requires a boolean (BOOL, yes or no, true or false, 1 or 0) statement to set the property of *hidden* in our label. The command is similar to setting a label's text value, *myLabel.text* = … and requires minimal work. The command is as follows:

```
myLabel.hidden = YES;
```

As for the value "YES", you can use any boolean value such as TRUE or FALSE, and 1 or 0. It's all developer preference, what ever floats your boat.

This is pretty much it for the coding part of our IBAction function, but we still need to go to the Interface Builder to add a button and associate it with the action.

Assigning Outlets and Actions in Interface Builder

Let's go to the storyboard file associated with the project to load Interface Builder.

You're going to want to scroll through the list presented at the bottom right portion of the window until you find the item titled "Button", not too far from the top. Drag and drop the item wherever you want on the same view with the existing label. You can change the text shown on the button from "Button" to whatever you want, or just leave it as is. Click the "View Controller" item with the orange-yellow icon in the mid left of the window. Then, click on the right arrow button in the top right of the window, scroll down it's associated list until you find the "Received Actions" section. In that category is where all of your declared IBActions hide for the winter. Click and drag from the name of your action, all the way to the button you just created. Once you let go, you will be presented with many options for the method of activating the action. Select the "Touch Up Inside" option, as that only activates the action once the user has touched, and let go of the button. There are some other self-explanatory options that you could use, but this is the most efficient (in my opinion).

Once you have completed all of the above steps, you are ready to build again with your preference of device!

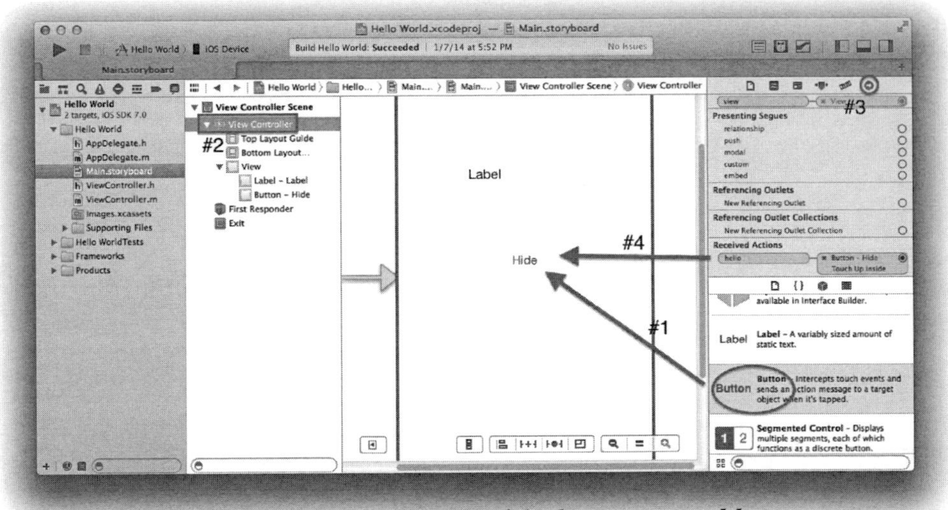

Figure 4.1 - Diagram with the steps to add a button and assign it's action

4.3 Hello UIAlertView

Alerts are a vital part of the app ecosystem. Most apps have some version of the UIAlertView in them some way or another. Whether it's asking you to review their app(s), download some other app, give you some sort of error, or even encourage you to buy some in-app purchases. Our approach won't ask for any money, but simply broadcast a nice message.

You may be catching on to the fact that we are building upon the original Hello World that we created back in Chapter Three, and are continuing to use the functions that we made earlier. This is all part the plan to make you retain this code well, as it is a vital part of app development.

We'll want to get started in the first section of the header file (where the string declaration took place) to declare a UIAlertView such as:

```
UIAlertView *alert;
```

It's that simple! We don't want or need to include *IBOutlet* before *UIAlertView* as there is no such function in Interface Builder, all alert views are presented through code. Now head over to the method file to finish this off!

Once the corresponding method file has loaded, find the action function that we created. We are going to add our alert view method in there, so that whenever the action is called, the alert appears.

The skeleton code for any alert view is:

```
alert = [[UIAlertView alloc] initWithTitle:
(NSString *)message:(NSString *) delegate:
(id)cancelButtonTitle:(NSString *)
otherButtonTitles:(NSString *), ..., nil];
[alert show];
```

You may notice that whenever you are typing any Apple-developed code, the skeleton code appears on it's own! You can take advantage of this by either pressing the return key, or by pressing the right arrow key (▶).

The alert view code starts by allocating (*alloc*) memory (RAM) to *UIAlertView* and is followed by defining the contents and properties of the alert. The first property is the title field, which is defined by a string. For this we will put the string *myString* which we already defined earlier. The next property is the "message", or body, of the alert. Here, you can either reuse *myString* or create your own message by typing @"" and filling in the quotations with your desired text. Next, we have the "delegate", or owner, of the alert. This belongs to the group of files that created it, which in programming, would be *self*. Fill the "delegate" property with *self*. The cancel button is the button that when pressed, dismisses the alert (makes it go bye-bye). You simply need the give a string to the button, either use *myLabel* or @“”. The “other” section is where you can optionally put in the title of however many buttons you want that have their own, defined actions. Since we only want to show the text “Hello World” in our alert, we can give this option the value *nil.* Even though *nil* automatically follows the “other” property, *nil* must always end the command. To compensate for this, the “other” property is followed by *nil, nil.* The final part of presenting an alert is by typing the command *[alert show];*. This does what it looks like, shows off the new function.

That’s all for alert views! Here’s what the finished action function looks like:

```
-(IBAction)hello {

    myLabel.hidden = YES;
    alert = [[UIAlertView alloc]
initWithTitle:myString message:@"Hi"
delegate:self cancelButtonTitle:@"Done"
otherButtonTitles:nil, nil];
[alert show];

}
```

Now it’s time to build! Build with your preferred device. Whenever you press the button assigned to your action, an alert will also appear now.

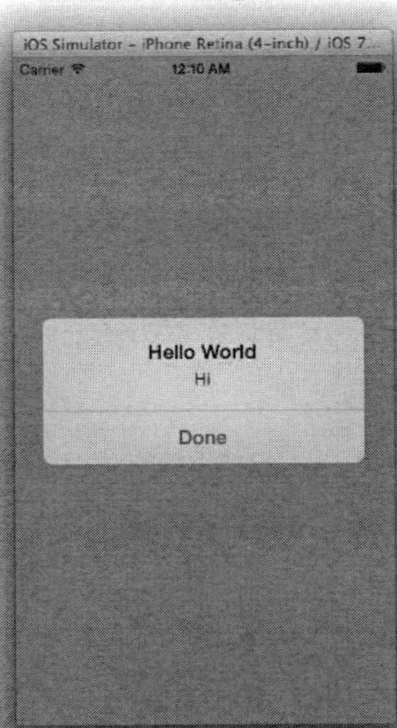

Figure 4.2 - Simulator presenting UIAlertView

4.4 - Hello UISwitch

The UISwitch item is typically used to change options or settings, but here we will use it to toggle the text's visibility.

The UISwitch, like most light switches, have two states, on or off. Light switches are real life boolean objects, they only have two states! In programming, these states are represented by 1 and 0, true and false, or yes and no.

What we want to do to the program, is add a function dedicated to the switch, which tells the program what to do whenever the switch is toggled. We also want the switch to be turned off whenever the action button, which hides the text, is pressed.

First, we have to declare a UISwitch in the header file. Since this is a tangible item, we must put *IBAction* before it. The product turns out to be:

```
IBOutlet UISwitch *mySwitch;
```

Now we go back to the method file. The first thing we need to do is in *viewDidLoad*. The newly declared switch needs to know what type of void function to listen to, so we must write some code that tells the switch just that:

```
 [mySwitch addTarget:self
action:@selector(switchChange:)
forControlEvents:UIControlEventValueChanged];
```

Now for the analysis! The code starts off by stating what item it is changing, *mySwitch*, and then says where the switch is to be found, *self*. The "action" command needs to be assigned a function, so we put *@selector* followed by the name that we want in parenthesis, *switchChange*, and that is followed by *:* and a closed parenthesis. The final section asks when the *switchChange:* function should be called, which is whenever the control's value changes.

We now need to create the *switchChange:* function. Even though not declared in the header file, it's still possible to create a void function in the method file alone. Although, there will not be an autofill feature for functions only defined in the method file. The name of our function will be the same as the name of the *@selector* name that we gave, except we will add *(id)sender* following *:* because the new function must have a sender to return to. The function will look as follows:

```
-(void)switchChange:(id)sender {
}
```

The code will feature an if-else statement which is a piece of code that, if true, runs a select group of code, and if it doesn't, it runs the else statement. For this if-else, if the switch is turned on by tapping the switch, then the label will not be hidden. The else is to make the label hidden.

```
-(void)switchChange:(id)sender {

    if ([sender isOn]) {
        myLabel.hidden = NO;
    }
    else {
        myLabel.hidden = YES;
    }
}
```

Maybe you're thinking "What if the button is pressed, which hides the label, AND the switch is still on?!". It'd be weird if you thought the exact same thing as written here, but regardless, we need to make sure that this doesn't happen. In order to prevent it, in the action that you previously made, add this code anywhere in the action's code block:

```
[mySwitch setOn:NO animated:YES];
```

This code makes sure to turn off the switch and make the motion animated.

Now that the programming is done, go to the storyboard file so we can add a switch to the program. Once there, scroll down the item list to find “Switch”. Add it to the view, and go to the typical *View Controller » Outlets* section, and connect *mySwitch* to the actual switch.

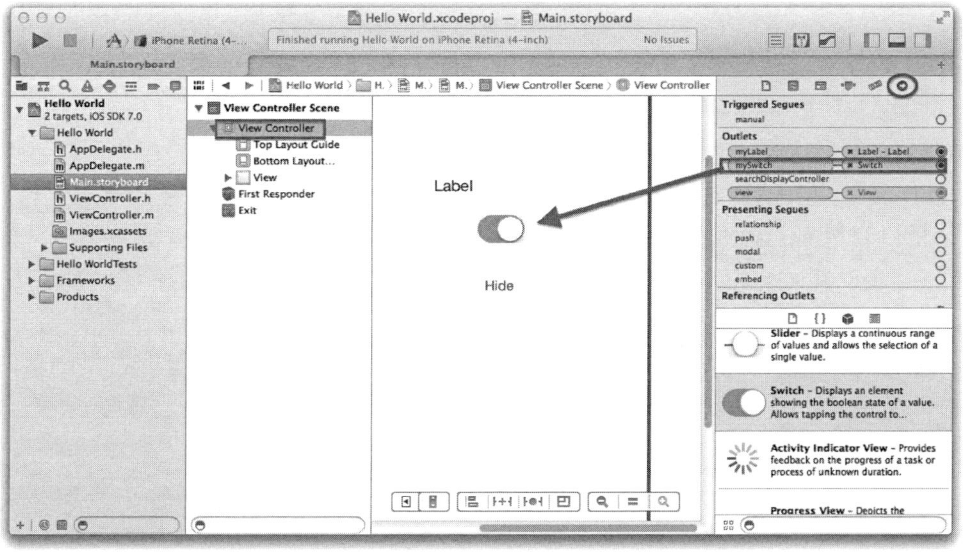

Figure 4.3 - UISwitch in Interface Builder

The program is now ready to run! In order to make the action work, simply click on the switch! I’d recommend pressing the button and using the switch a few times just to make sure that everything works properly. If there is anything wrong, just recheck your code with the code provided.

4.5 - Hello UIImage and UIImageView

“Now this is completely different than all this other crap I already learned” you may be thinking (hopefully without the “crap”). Well not quite, reader! There may be no strings, labels, or void functions involved, but adding images is just like all the other stuff we did! You declare stuff in the header file, then define them in the method file. There's only one stipulation, you have to import an image into Xcode… God forbid! It's not actually that hard to import anything into Xcode, you can just drag and drop, or do the old *File »* … option if you prefer. It's time to stop talkin' to talk, and start walkin' the walk!

First things first, download or find an image that you want to put in the app… done?! Grand, glorious, wonderful! Now either drag the file into the part of Xcode that has all of the files and folders, or click *File » Add Files to…* and select the file, to be presented with a screen that asks if you would like to "Copy items into destination group's folder". This is basically asking if you would like to make a copy of the file; I would recommend checking this box so that the project isn't ruined if you accidentally delete the file. Once you are done procrastinating, click "Finish" and the file will be added to the project.

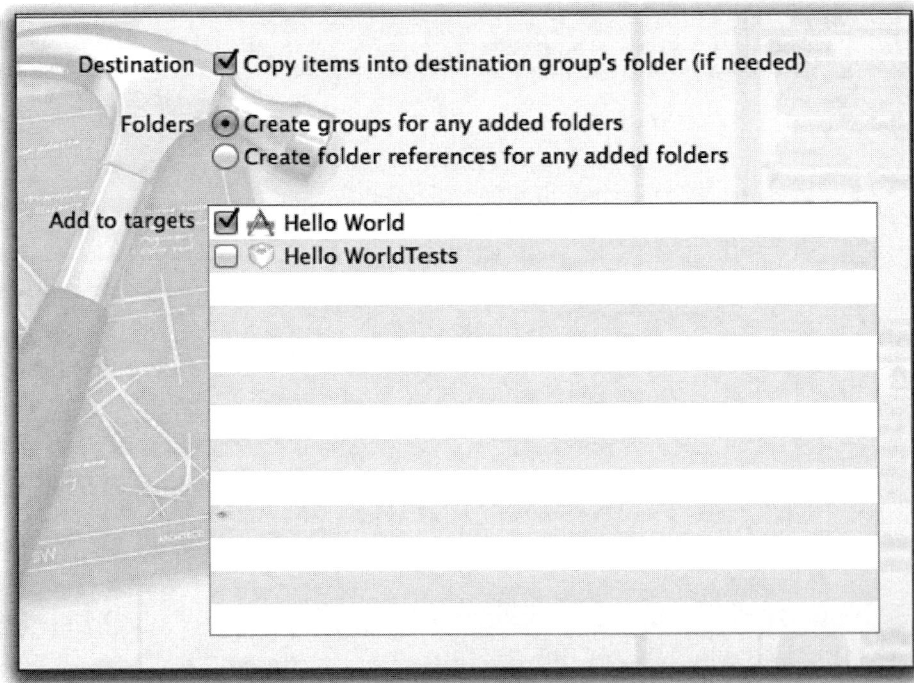

Figure 4.4 - Adding files to Xcode

Now that you have the image in Xcode, we can really get started! Go to the header file so that the image can be declared.

To make an image show, we need two different things, a UIImage, and a UIImageView. It's kind of annoying, but we just have to deal with it. UIImage declared as:

```
UIImage *myImage;
```

UIImageView is declared by following an outlet because UIImageView is the only image option in Interface Builder:

```
IBOutlet UIImageView *myImageView;
```

We're done here, and are ready to head to the method file. Since there is nothing to activate with images, all of the coding will take place inside of *viewDidLoad*. The first step is by defining what *myImage*'s image is. From there, we have to assign *myImage* to take care of *myImageView* by changing *myImageView*'s property from UIImageView to UIImage and specifying the name of the image. All of this leads to:

```
myImage = [UIImage
imageNamed:@"XcodeScreenSpapz017.png"];
myImageView.image = myImage;
```

The coding is already done, and you thought we were just getting started! We need to finish this baby off by going to Interface Builder.

First, we need to drag and drop an "Image View" from the list of items onto the view, wherever you want, however big you want. You then need to find the "Outlets" section and connect the *myImageView* outlet to the image view (notice I'm giving less hints per screenshot).

Once completed, run the app to make sure that the image appears. It's okay if the image isn't in the right dimensions, all that matters now is that the image actually shows up. If this is the case, it is possible that the image you imported is too big or really, really, really, small. Another problem could be that the file name and code representing the file don't match. Check that they match, and change the code if they don't match.

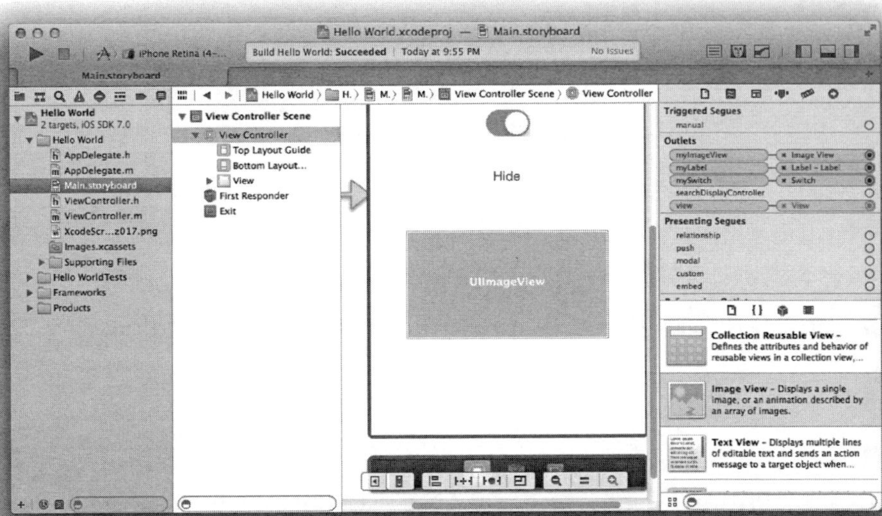

Figure 4.5 - UIImageView on a view

Ch. 4 Conclusion

You're probably glad to finally hear it, the Hello World project is OVER! The rest of the book is all new content (even through pretty much everything in this chapter was new)! It's been a wild ride with Hello World, but you won't have to see that phrase again unless you (lovingly) reread the book. As with the end of every chapter, here is the checklist of what you should know by now. We are just getting into coding, so don't feel bad if you still don't understand something, just go back and reread the section you don't understand.

- You can declare a void function ☐
- You can declare a variable ☐
- You know that switches are controlled by bool statements ☐

- You are DONE with the Hello World project ☐

Chapter 5

The Sound of Music

Creating a Simple Soundboard App

One of my early apps was a simple (and ugly) soundboard that had about forty different funny sound clips (some copyrighted... oops, I was young). It was actually kind of popular, which made this my first app that I was actually proud of! In this chapter, you'll make your own soundboard with whatever you want! In order to complete this project, we'll use and need:

- IBAction
- Void
- NSURL
- NSString
- Two to six mp3 and/or wav sound clips
- AVFoundation Framework

The goal of this project is to use a new-to-you framework, *AVFoundation*, to play sound files that have been added to the project bundle by you. The design will be a simple "Single View" project that has various buttons that call a void function which plays the corresponding sound clip, and one button that stops the sound clip's playback.

Want to get started? First, I'd recommend finding two to six mp3 and/or wav files that you'll want to play in the app. Don't worry about copyright on the sound clips unless you plan on actually releasing the app on the App Store, free or paid, Apple takes copyright very seriously!

5.1 Setting Up the Project

The creation of this project is very similar to that of the Hello World project. We're going to want to start by launching Xcode, and clicking "Create a new Xcode project". Just like last time, this application will run off of the "Single View Application" template that Xcode provides for us. Select it, then hit "Next". On this

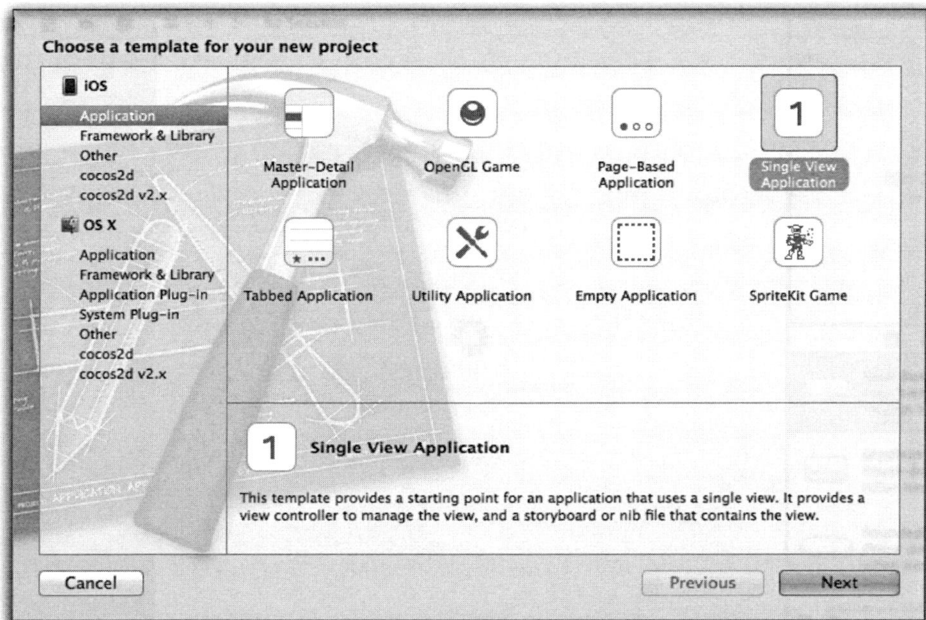

Figure 5.1 - Single View Application option

window, you're going to need to name the project. Choose whatever "Product Name" you want (as long as you don't reuse a name), and the "Organization Name" and "Company Identifier" should now be defaulted as the values of the most recent project. Again with this project, we want the "Devices" value to be set at "iPhone" so that the app is only optimized for the iPhone. Don't worry, we'll get to the iPad later (it's too much to handle right now)! After you're done with this screen, it'll ask where you want to save the new .xcodeproj file. Choose wherever, although, I would recommend saving it in a place where you could easily find it. Once finished, click "Create" so we can get started!

First things first, go to your header file and add an open and closed curly brackets after *UIViewController*.

```
@interface ViewController : UIViewController {

}
@end
```

Now that the header file has an extra section, we want to turn around and do something different in the header! Since this app will become a soundboard, we need to import a framework that supports the playback of sound files. This means we have to import the required framework into the header file. It's only required to be imported in the header file because the method file imports the header file and all of the things that the header imports. What we will be using is the *AVFoundation* framework. The code for importing looks like:

```
#import <AVFoundation/AVFoundation.h>
```

Importing a Framework

You may be wondering why importing frameworks is required, and not just built into the main framework. There is a method for this madness! It's due to the fact that every line of code takes up some temporary RAM and some temporary storage space. If every framework was combined, the app would be slow to launch and function, along with the fact that the user's device may not have either enough RAM or storage space for the app to function at all, and could lead to a total device crash. This doesn't mean that you can't use a bunch of frameworks in your app, it's just that there are a TON of frameworks, and I doubt you'll ever use all of them because some are just pointless for many developers.

Back to *AVFoundation*! The first of two parts of the import command is stating that what will be imported is from the *AVFoundation* framework, and the second part details which exact part you are importing. By retyping the framework name followed by *.h*, all of the files in the framework are imported. It is possible to only import one file from a framework to cut down on the app's file size.

In order for the program to recognize the framework, we first need to manually import it into the program. To do this, click the project name and icon at the top left, then the "Build Phases" category towards the right, open the section titled "Link Binary With Libraries", and finally press the "+" button to be presented with all of the frameworks available.

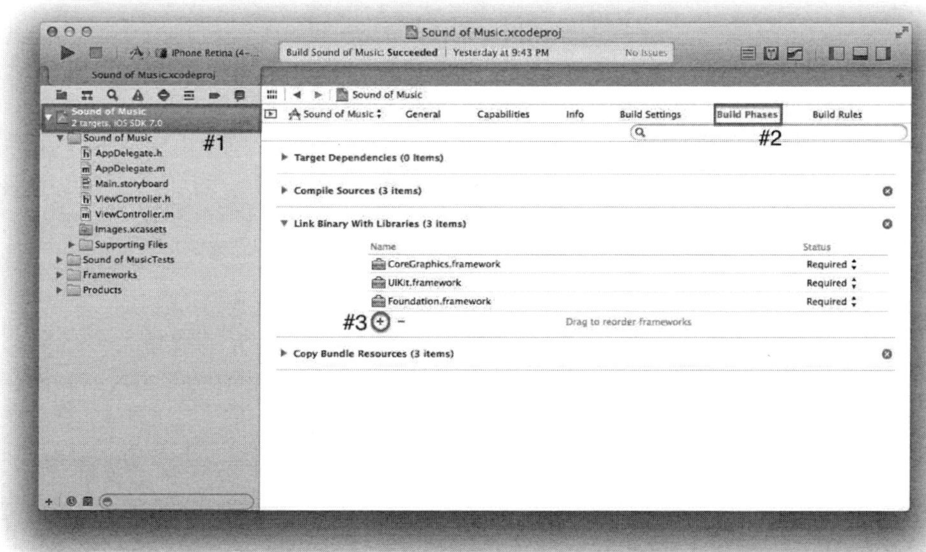

Figure 5.2 - Steps to adding a framework

On that window, find, select, and add the framework titled, "AVFoundation.framework" and press "Add" for the framework to be finally imported into the project! Let's return to the header file so that we can add some items and functions.

Once in the header file, we need to add the item which will play the sound files, *AVAudioPlayer*. This is just a simple declaration in the first declaration section of the header:

```
AVAudioPlayer *player;
```

Next we're going to declare some strings that will specify the sound files' names and the file type. This, like the *AVAudioPlayer* is a simple declaration in the same location:

```
NSString *currentSong;
NSString *currentFileType;
```

The final things we need to declare are functions. We're going to need one void function that will be the command to play the sound file, and three to seven action functions; one action for stopping the file, and two to six actions (feel free to go above and beyond with more than six actions) for setting up the various sound files.

```
-(void)playSound;
-(IBAction)stop;
-(IBAction)sound1;
-(IBAction)sound2;
-(IBAction)sound3;
-(IBAction)sound4;
```

In my case, I'm going to be using four different sounds, the stop playing action, and the function that plays sound files. The reason I'm not using six sound actions is because I want you to be able to transfer what I teach in the first four sound actions and do the next two (plus) actions all by yourself. Half the battle of learning code is becoming independent by retaining and catching onto code patterns. It's time to move onto the method file, but first check my code to make sure it matches yours:

```
#import <UIKit/UIKit.h>
#import <AVFoundation/AVFoundation.h>

@interface ViewController : UIViewController {

    AVAudioPlayer *player;
    NSString *currentSong;
    NSString *currentFileType;

}

-(void)playSound;
-(IBAction)stop;
-(IBAction)sound1;
-(IBAction)sound2;
-(IBAction)sound3;
-(IBAction)sound4;

@end
```

Once the method file has loaded, you may notice that you have a few yellow warning signs. “Why is this? I followed all of Chip’s code so why does this thing appear?” you may be thinking (or even “Da*n, s*it, F**K, and/or son of a pastrami… whatever floats your boat). The answer is simple, Xcode wants to help you out by letting you know that you have items and/or functions that are lying around doing nothing, and most importantly, wasting memory and storage. This warning can become VERY annoying, but the solution is also simple. Add all of the unused functions in the method file. Let’s fix that now! Add all of your declared functions and actions into the main section of the method file as such:

```
-(IBAction)stop {

}
```

Do this step for each action and function while replacing the type of function and name where needed. Once completed, your code should look similar to mine, with variations permitted in name, order, and quantity:

```
-(void)playSound {

}

-(IBAction)stop {

}

-(IBAction)sound1 {

}

-(IBAction)sound2 {

}

-(IBAction)sound3 {

}

-(IBAction)sound4 {

}
```

If you get any errors marked in red, make sure that this code is past the first *@end,* before the second (last) *@end,* and is not accidentally part of *viewDidLoad,* or any of the other default, template functions that Xcode includes.

5.2 It's All About Content

Now that the basic setup of the project has been completed, we can move onto filling in all these blank functions with some brand spanking new code. Before we do anything too strenuous, you need to import the actual sound files that you want to use in the project.

Importing a sound file, or any file, into Xcode is the same as when we imported an image, back in the previous chapter. In case you forgot (which is understandable), all you need to do is drag and drop the files you want into the right side bar, and follow the onscreen prompt. Or, in Xcode, click *File » Add Files to…*, select the files you want add, follow the onscreen prompt, and make sure to select the option that creates a copy of the file for use in the project. Once added, we can continue.

It's vital to your understanding of the topic to start with the actions that are assigned to the sounds as it's easy to get lost otherwise. So, go to your first sound action. The action that we want to achieve by pressing a sound button is to assign the declared strings a value that matches that of the sound clip you want to play, and then call upon the void function to run. What this looks like is:

```
-(IBAction)sound1 {

    currentSong = @"soundFileName1";
    currentFileType = @"mp3";
    [self playSound];

}
```

The first action performed is assigning *currentSong* to the name of the sound clip imported earlier. In my case, it is simply "soundFileName1". The next step is assigning the string, *currentFileType,* to the format of the imported sound clip, either wav or mp3. The final line is a *self* command which tells the program to run the function named *playSound.*

Rinse and Repeat

A lot of programming is all about repetitiveness. When programming, you form habits (Bostonian's version of Hobbit), both good and bad. The point of this book is not to be a "copy, paste, and go" book, but a technical book that has code that is meant to not only be copied, but to be thought about while typing. It's your choice whether you simply "copy and go" or think and learn along the way so that your journey may be long, but very worthwhile. Throughout this book, there will be many points where you feel like copying and pasting code, but I plead that you stick it through so that you can move onto greater things upon completion.

Going back to the action that we just made, we're going to repeat that process with the rest of the sound actions, only changing the *currentSong* and *currentFileType* values along the way. The product should like something like:

```
-(IBAction)sound1 {

    currentSong = @"soundFileName1";
    currentFileType = @"mp3";
    [self playSound];

}

-(IBAction)sound2 {

    currentSong = @"soundFileName2";
    currentFileType = @"mp3";
    [self playSound];

}

-(IBAction)sound3 {

    currentSong = @"soundFileName1";
    currentFileType = @"mp3";
    [self playSound];

}
```

```
-(IBAction)sound4 {

    currentSong = @"soundFileName1";
    currentFileType = @"wav";
    [self playSound];

}
```

Making Sweet Music

Now that we've typed so much about the *playSound* function, it's finally time to make it do some work. The goal is to make the function play the sound file that is currently tied to the strings *currentSong* and *currentFileType*. This will be done by creating a URL in the function that is readable by the audio player, and is equal to our declared strings. We then have to assign the player to the current value of the URL, and then tell it to play. Now go to the *playSound* function. Let's start out one line at a time:

```
NSURL *url = [NSURL fileURLWithPath:[[NSBundle
mainBundle]pathForResource:currentSong
ofType:currentFileType]];
```

The first thing you might notice is that we are declaring a URL directly in the method file… what tyranny! This is actually called a local variable. Local variables are variables initialized directly inside of a function or piece of code. The point of this type of variable is to save memory by only creating the variable when needed. It's useful whenever you only need a variable once or twice in only one function or action. Please be aware, local variables can only be used within the curly brackets that they initialized in.

The next part of the code is defining where this new variable is to get it's information. The *mainBundle* referred to in the code, is the "bundle" of files that make up the project as a whole, this includes the files that are imported. Next is the *pathForResource:* which we want to be set to the *currentSong* string, which is always equal to the last button touched. The *ofType* property is to be set to *currentFileType*, which provides the file format of the latest button tapped. The next lines of code:

```
NSError *error;
    player = [[AVAudioPlayer alloc]
initWithContentsOfURL:url error:&error];
```

Because the *AVAudioPlayer* class requires there to be a backup if an error occurs, it is required to simply initialize an error locally before you do so with the audio player. The next block of code is telling what the *AVAudioPlayer* named *player* to do. In this case, we must first allocate memory with *alloc*, and then set the URL contents to be equal to the locally initialized URL, *url*. The last part of this code is setting the in-case-of error equal to the locally initialized, *error*, which is preceded with an ampersand (&), which in programming, points the nil-ed *error* to an error unique to the audio player. The last line of code:

```
[player play];
```

I hope this is pretty self explanatory… the *AVAudioPlayer*, *player*, is told to play. Now that the *playSound* function does all that we want it to do, it's time to move onto the *stop* action.

The action, *stop*, should also be self explanatory, we want the *AVAudioPlayer* named *player* to stop playing (all this alliteration is awesome)! The code for this is just like getting *player* to play, except the opposite. Go ahead and try to guess the code (insert "Jeopardy" music)! Got it, great! Don't got it, that's okay also because I'm here to teach, and you to learn. The code is as follows:

```
[player stop];
```

Wow! How simple is that!? This is all the coding that we need to do for this project… congrats! Before continuing, I recommend that you compare my method file to yours, and fix any mistakes if necessary.

```
#import "ViewController.h"

@interface ViewController ()

@end

@implementation ViewController

-(void)playSound {

    NSURL *url = [NSURL fileURLWithPath:
[[NSBundle mainBundle]

pathForResource:currentSong

ofType:currentFileType]];
    NSError *error;
    player = [[AVAudioPlayer alloc]
initWithContentsOfURL:url error:&error];
    [player play];

}

-(IBAction)stop {

    [player stop];

}

-(IBAction)sound1 {

    currentSong = @"soundFileName1";
    currentFileType = @"mp3";
    [self playSound];

}

-(IBAction)sound2 {

    currentSong = @"soundFileName2";
    currentFileType = @"wav";
```

```
    [self playSound];

}

-(IBAction)sound3 {

    currentSong = @"soundFileName3";
    currentFileType = @"mp3";
    [self playSound];

}

-(IBAction)sound4 {

    currentSong = @"soundFileName4";
    currentFileType = @"wav";
    [self playSound];

}

- (void)viewDidLoad
{
    [super viewDidLoad];
     // Do any additional setup after loading the
view, typically from a nib.
}

- (void)didReceiveMemoryWarning
{
    [super didReceiveMemoryWarning];
    // Dispose of any resources that can be
recreated.
}

@end
```

5.3 Doing Some Interface Building

Tis the time to go the project's storyboard file. Once loaded, drag as many buttons as you need from the lower right hand list to the view. Place them wherever you want on the view, as long as they are still visible. You can also name them however you want, although, I would recommend giving them a memorable or literal name. Once completed, click on the yellow-orange "View Controller" item in the mid-top left, then the right arrow item in the top right. Next, scroll to "Received Actions" section and drag each item to it's corresponding button while choosing the option "Touch Up Inside" or "Touch

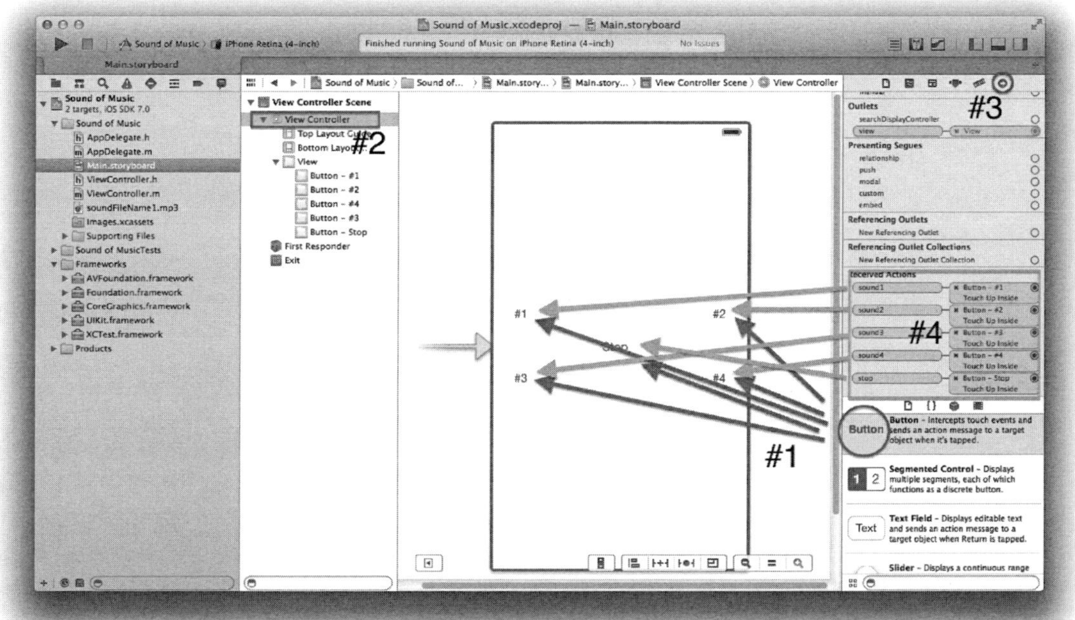

Figure 5.3 - The action conglomeration/mess

Down" (not football).

Yay! Your first non text-based app is finished. You know what that means, building and testing! When built on an iPhone or on the iOS Simulator, the app should look exactly as designed in Interface Builder. Whenever you press a button containing a sound, the sound will play within a couple seconds, and if you press the button that calls the *stop* function, the player will stop playing.

Troubleshooting

If the sound fails to play on a physical device, first check that your device is not on vibrate, and has the volume level up. If this isn't the problem, check to make sure your connections in Interface Builder are correct. If it still fails to work, recheck your code so that it matches mine.

If the sound fails to play on the iOS Simulator, check that your Interface Builder connections are correct. If it still fails, recheck your code so that it matches mine.

If the app fails to launch on a physical device, check that you have valid provisioning profiles selected, and that you have no errors (at all) in the program.

If the app fails to launch in iOS Simulator, check that you don't have any errors (at all) in the program.

CH. 5 CONCLUSION

Congratulations, you now have an app worthy enough to show off to your friends (I was lying last time I said this). This was actually a pretty fun project for me, as well as for you, I hope! We have finally broken the boring text-based app barriers, and we are charging from one new thing to the other many times per chapter. I bet that by now you could write little bit of code on your own, or at least can grasp the concept of what you would need to do to make an app! As per usual, here's the "chapterly" review checklist.

- You know how to import a framework ☐
- You know what a local variable is ☐
- You can at least initialize or declare a function ☐
- You can navigate Interface Builder better than last time ☐

Chapter 6

Tweets and Giggles

Creating a Joke Generator While Using the Social Framework

A common occurrence in this book is the fact that I use the apps that I have previously made and published as lessons. I know that in my experiences, each and every app that I worked on I learned lots of new things. This is the general goal of this book, to teach you the fun basics of each corner of iOS app development so that you may put together the code in your own ingenious ways.

In this chapter, you'll be learning about the new ways of displaying text, creating a random event, about Apple's pretty recent framework simply titled *Social,* and the *MessageUI* framework. We'll create a random joke generator that has the ability to post the joke on the social networks Twitter and Facebook, as well as learn how to send the joke via iMessage.

In this chapter we'll need and use:

- *Social* framework
- *MessageUI* framework
- Switch case statement
- UITextView
- IBAction
- UIButton

6.1 Getting All Setup (Again)

Just like the past couple chapters, we're going to create a new project and use the "Single View" template (this template can handle pretty much anything!). Feel free to name the project as you please, remember… you have the freedom to type!

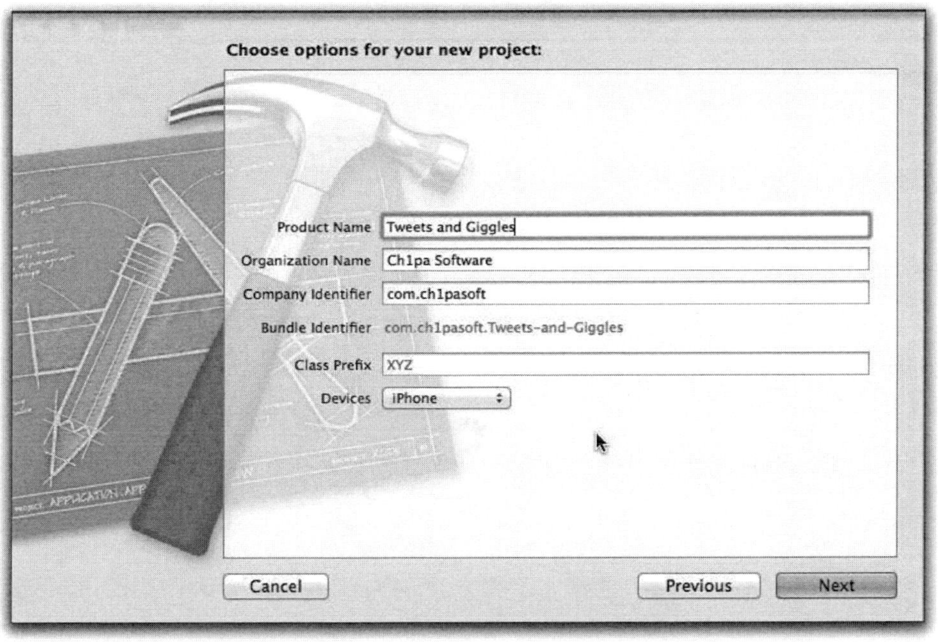

Figure 6.1 - Creating a project

Once named, click "Next", select a reasonable destination to save, and press "Create".

Ah… back to square one with a blank slate. Seeing an uncluttered and empty file of code can both be intimidating and calming, you don't have to worry about errors, but you have to worry about where to start.

We're going to start a little differently than last time by importing the frameworks, *Social* and *MessageUI*, first. If you aren't already there, go to the main project hub by clicking your project's name in the left side bar. Then, near the top-middle, click the "Build Phases" option, next the "Link Binaries With Libraries" row, and click the plus button at the bottom of the row. You will then be asked to select which framework you would like to import; scroll to the bottom of the list and add the *Social.framework* file and then the *MessageUI.framework* file. You can exit this section now and head over

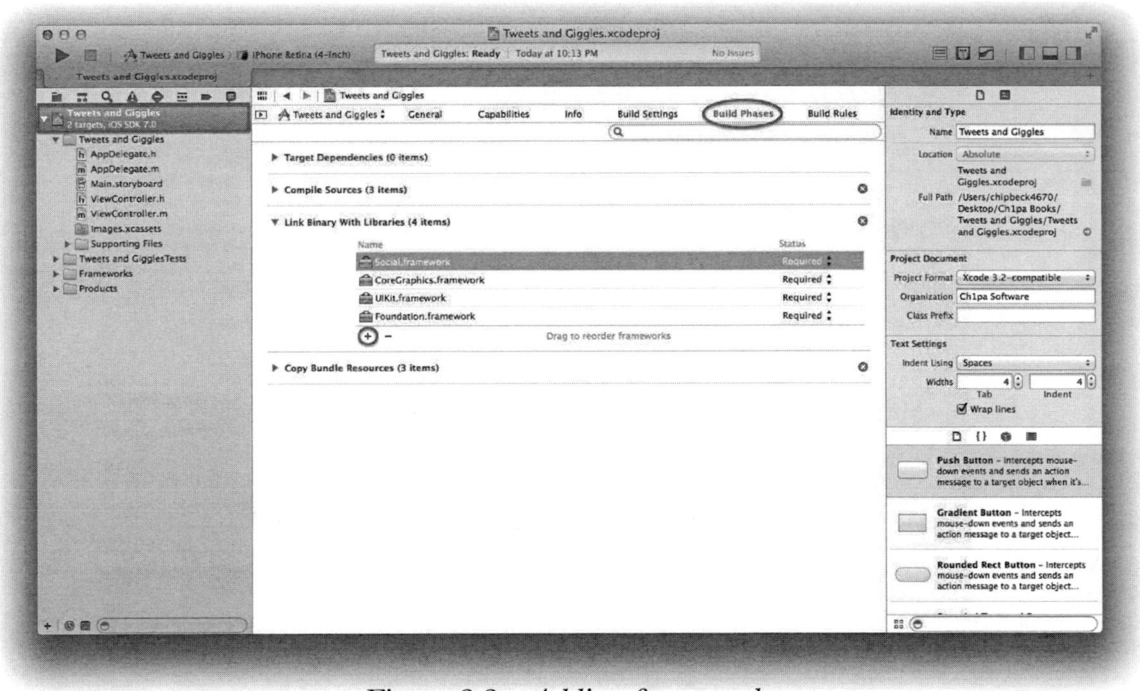

Figure 6.2 - Adding frameworks

to the header file.

In the header file, we are first going to import the whole *Social.framework and MessageUI.framework.* This is done the same as the last time, by using the code:

```
#import <Social/Social.h>
#import <MessageUI/MessageUI.h>
```

We then want to divide the header into two main parts with the curly brackets. In the first section, we're going to initialize a NSString and a UITextView:

```
NSString *jokeString;
IBOutlet UITextView *jokeView;
```

Now in the second section, we are going to need an action function for generating a joke, an action for tweeting, an action for posting to Facebook, and an action for sending an iMessage. Simply create three action functions:

```
-(IBAction)generate;
-(IBAction)tweet;
-(IBAction)post;
-(IBAction)message;
```

Before moving on, we need to assign our header file to the delegates *MFMessageComposeViewControllerDelegate* and *UINavigationControllerDelegate* (don't ask me about the navigation one, I don't even know). This is placed before the opening curly brackets such as:

```
@interface ViewController : UIViewController
<MFMessageComposeViewControllerDelegate,
UINavigationControllerDelegate> {
```

Now let's go to the method file and set it up to get rid of those pesky warnings. We're going to want to declare an action function for each action we initialized in the header. Place the code after the line containing *@implementation ViewController* and before the second *@end* to avoid errors. Your code should look something like:

```
-(IBAction)generate {

}

-(IBAction)tweet {

}

-(IBAction)post {

}
```

```
-(IBAction)message {

}
```

6.2 Info Had Me Like

As cute as the code we already have may look, it still needs to be filled up in order for it to do stuff! The first place we are going to want to fill up is the joke generator action. For this generator to work, it'll have to use a switch case statement run by a randomizer. "Whoa, whoa, slow down, what is a switch case statement exactly?" Good question reader! A switch case statement is a statement similar to an if-else statement, which determines what code to run via the value of an integer. For example: if number = 1, do this, if number = 2, do that. In my opinion, I don't find switch case statements too useful in doing anything unique, although, they do save some time on coding a statement that has many possibilities (such as a joke generator).

Coding Switch Case Statements

A switch case statement must be based off of an integer, so we must start out by locally initializing an integer, but not just any integer... actually yes... just any integer, a randomized integer. To define a random integer, we must use a randomizer. There are plenty of randomizers out there, but the one that works best (arguable) on iOS is the generator, *arc4random*. The next step is defining the amount of possible statements. The initialization looks like this when ten is the max:

```
int text = arc4random() % 10;
```

Now, every time that the action *generate* is tapped, the integer, *text*, is equal to a random integer between zero and nine. "Wait what!" says you. I guess I forgot to mention that random integer generators start at zero since zero is technically an integer! The number ten comes from the number of integers that can possibly be created, and zero through nine makes ten integers. Sorry for that little interruption!

The next portion is the big kahuna! It's the part where we define every case statement possible… how fun (he said sarcastically). The actual statement begins with *switch* followed by an integer in parenthesis. The statement is then followed with an open curly bracket and then the possible statements (a.k.a. cases). Each case begins with a colon, and is finished with *break;*. The entire switch statement is finished with something called *default:*, in which is always run whenever finished running the case statement. All together, the switch-case statement looks like:

```
-(IBAction)generate {

    int text = arc4random() % 10;
    switch (text) {

            case 0:
            break;
            case 1:
            break;
            case 2:
            break;
            case 3:
            break;
            case 4:
            break;
            case 5:
            break;
            case 6:
            break;
            case 7:
            break;
```

```
            case 8:
            break;
            case 9:
            break;

    default:
    break;

    }

}
```

Currently, this code sadly does nothing at all. We need to add the code we want in-between the *case #:* and *break;*. The code that goes here is the code that sets our string to whatever joke you want. It's fairly simple:

```
jokeString = @"This is a joke... ha";
```

Since I figured that using a UITextView would be more difficult and would require more coding, I was all for it! To set the UITextView equal to the string, you must type:

```
jokeView.text = jokeString;
```

The above code goes into the section of the switch case titled *default:* because this is the "default" code that must happen after a case statement in order for the UITextView to actually display and update it's contents.

It's now time to fill up ten cases with jokes! Use the same code above that changes the value of *jokeString*, but change the joke of course because the app would be quite pointless without it! Make sure that there is a *break;* at the end of each statement:

```
-(IBAction)generate {

    int text = arc4random() % 10;
    switch (text) {

            case 0:
```

```
            jokeString = @"This is a joke... ha";
            break;
            case 1:
            jokeString = @"What is the first
programming language you learn when studying
computer science? Profanity.";
            break;
            case 2:
            jokeString = @"Look at Chip!";
            break;
            case 3:
            jokeString = @"Why is he so weird?";
            break;
            case 4:
            jokeString = @"So I heard Chip makes
apps!";
            break;
            case 5:
            jokeString = @"Ya, and I heard he
wrote a book too!";
            break;
            case 6:
            jokeString = @"Wow, he must be the
bee's knees.";
            break;
            case 7:
            jokeString = @"What?! Are you from
the freakin' 20's!";
            break;
            case 8:
            jokeString = @"No! I was just
joking.";
            break;
            case 9:
            jokeString = @"You better have
been...";
            break;

            default:
            jokeView.text = jokeString;
            break;
```

```
    }
}
```

Just to recap, in order update the visible UITextView's value, we must set it's text property to become the same as the value of the string that changes every time the *generate* action is performed. It's important that this line of code goes in the *default:* section, because if it didn't, the UITextView would never update to the latest joke received by the string.

6.3 Becoming Social

Back when iOS 5 was announced, Apple revealed that iOS would be deep-integrated with Twitter. This meant that developers could integrate Twitter into their apps without even having to go through Twitter and it's complicated SDK (software development kit). Tweeting was now only a matter of including a couple lines of code! Then came iOS 6 with Facebook and Sina Weibo (Chinese) deep-integration which threw out the Twitter framework and made a unified Social framework. THEN, with iOS 7 came the Chinese social network, Tencent Weibo. iOS 7 also brought Flickr and Vimeo deep-integration, although, at the time of writing, Flickr and Vimeo have yet to be third-party (us) accessible as part of the Social framework. In this book, we're only going to focus on Facebook and Twitter

Posting a Post (Facebook)

Facebook, I'm sure you're familiar with it. Although not very popular with teens, Facebook still has billions of members and won't be forgotten any time soon. Facebook has the option of posting text, images, and links. All of these items are allowed to be posted via the Social framework! Let's start coding!

We need to start out by locally declaring a Social framework view controller, and assign it to the service type named Facebook:

```
SLComposeViewController *poster =
[SLComposeViewController
composeViewControllerForServiceType:SLServiceType
Facebook];
```

Next we need to set our Facebook post's text equal to the current value of *jokeString*. This cannot be done via our standard *name.text = string* command because the Social framework needs to be able to differentiate links and plain text, as there is a separate field for URL links. So… the code is as follows:

```
[poster setInitialText:jokeString];
```

Feel free to add a URL to the tweet by typing

```
[poster addURL:[NSURL
URLWithString:@"Ch1paSoft.com"]];
```

The next piece of code tells the program (*self*) to show the Facebook post template with the attached properties that we have already set such as the URL and text. This piece of code is not exclusive to the Social framework, so the final property of this code will be set to *nil*. The code is:

```
[self presentViewController:poster animated:YES
completion:nil];
```

Even though we set the *completion:* property to *nil*, we still want a completion handler to let the user know that the post succeeded via UIAlertView. The code is much more involved for this, so I will show you the code before I explain it:

```
poster.completionHandler =
^(SLComposeViewControllerResult result) {
```

```
        if (result ==
SLComposeViewControllerResultDone) {
            UIAlertView *alert = [[UIAlertView
alloc] initWithTitle:@"Congrats" message:@"You
tweeted dat joke" delegate:self
cancelButtonTitle:@"Fo shizzle"
otherButtonTitles:nil, nil];
            [alert show];
        }
        [self dismissViewControllerAnimated:YES
completion:nil];

    };
```

Believe it or not, all of this code is one command statement, even though it includes curly brackets and an if statement! It begins off simple, by stating it wants to set a value for the property *completionHandler*. Then, it gets complicated. Since there is so much to do whenever the post has been sent, we need to create a code block focused on the result of the post, a success, cancellation, or failure, and the result is stored in *result*. We open the code block with an if-else statement that runs if *result* == *SLComposeViewControllerResultDone*. Side note, when comparing items you use ==, and when setting items equal to something you use =. Inside of the if-else statement, there is a UIAlertView which lets the user know that whatever was contained in the *SLComposeViewController* was successfully sent and posted to Facebook. Since we don't need an else for this statement, we finish with the dismissal of the Facebook post composition. We then close the code block with a closing curly bracket, and then finish it off with a semicolon because this code is all part of a single command started above by *poster.completionHandler*.

Wow! That's all for posting to Facebook, so make sure *-(IBAction)post* is all closed up with no errors, and we can move on.

```
-(IBAction)post {
```

```
    SLComposeViewController *poster =
[SLComposeViewController
composeViewControllerForServiceType:SLServiceType
Facebook];
    [poster setInitialText:jokeString];
    [poster addURL:[NSURL URLWithString:@""]];
    [self presentViewController:poster
animated:YES completion:nil];
    poster.completionHandler =
^(SLComposeViewControllerResult result) {

        if (result ==
SLComposeViewControllerResultDone) {
            UIAlertView *alert = [[UIAlertView
alloc] initWithTitle:@"" message:@""
delegate:self cancelButtonTitle:@""
otherButtonTitles:nil, nil];
            [alert show];
        }
        [self dismissViewControllerAnimated:YES
completion:nil];

    };

}
```

Tweeting a Tweet in the Twittersphere

First of all, yes, Twittersphere is an actual word according to the New Oxford American Dictionary!

Twitter has undoubtedly gained tremendous growth in the past couple years (with the help of Twitter integration in iOS 5), whether you like it or not (I personally don't), so I feel it is an important topic to teach, and to teach correctly. As you may know, Twitter has one truly defining feature, it's one hundred-forty character limit messages known as tweets (now a word with a new definition). This is why

Twitter integration must be taught correctly, to avoid errors and other unexpected results.

The code is very similar to posting to Facebook, the only differences are changing *SLServiceTypeFacebook* to *SLServiceTypeTwitter*, and adding an if-else statement that prevents jokes longer than 140 characters from being able to be sent, and then returned a failure. If the joke happens to exceed one hundred-forty characters, we'll simply place a UIAlertView to appear explaining the issue. The code for detecting string length is as follows:

```
if (jokeString.length <= 140) {

}
```

The code that we used for the Facebook post goes inside of the if statement's code block to prevent violators of the one hundred-forty character rule to run this code. Instead, the violators will run this:

```
else if (jokeString.length > 140) {

}
```

I do realize that I overdid this last statement as a number must either be greater than, equal to, or less than one hundred-forty, but I wanted to emphasize that else statements can also contain ifs.

The contents of the else statement is simply a UIAlertView that explains the problem:

```
UIAlertView *alert = [[UIAlertView alloc]
initWithTitle:@"Whoops" message:@"The joke is too
long!" delegate:self cancelButtonTitle:@"KK"
otherButtonTitles:nil, nil];

[alert show];
```

This is pretty much it for the differences between tweeting (also a word with a new definition) to Twitter via programming and posting to Facebook. In case something terrible happened to the Facebook code, or you have errors in the Twitter code, here is the finished *tweet* action:

```
-(IBAction)tweet {

    if (jokeString.length < 140) {
        SLComposeViewController *tweeter =
[SLComposeViewController
composeViewControllerForServiceType:SLServiceType
Facebook];
    [tweeter setInitialText:jokeString];
    [self presentViewController:tweeter
animated:YES completion:nil];
    tweeter.completionHandler =
^(SLComposeViewControllerResult result) {

        if (result ==
SLComposeViewControllerResultDone) {
            UIAlertView *alert = [[UIAlertView
alloc] initWithTitle:@"Congrats" message:@“You
posted to the Twitters" delegate:self
cancelButtonTitle:@"Fab" otherButtonTitles:nil,
nil];
            [alert show];
        }
        [self dismissViewControllerAnimated:YES
completion:nil];
    };

    }

    else {

        UIAlertView *alert = [[UIAlertView alloc]
initWithTitle:@“Er-r“ message:@"Too long biatch"
delegate:self cancelButtonTitle:@“Sucks to suck"
otherButtonTitles:nil, nil];
        [alert show];
    }
}
```

6.4 Messaging the Crap Out of Jokes

Say you love a joke you see and don't want to share it with the whole world, but just with one or two friends. What do you do? Your friends don't go on Facebook anymore, and hate Twitter, so… you text it! This is where the MessagesUI framework comes into play. MessagesUI allows the developer to either enable emailing or text messaging in the app under the end user's discretion. Let's focus on text messaging now, but don't fret (odd word) because emailing will come up in a later chapter. Go to your action function for messaging so we can get started!

Sadly, the MessageUI framework is totally different than the Social framework, so there is no reusing of that code. But, just like before, we do want to start with a local initialization of the composer, and in this case it is *MFMessageComposeViewController* (even longer than before!):

```
MFMessageComposeViewController *messageComposer =
[[MFMessageComposeViewController alloc] init];
```

Next, we want to set the body of the message along with who receives it. This is relatively simple, the message will equal *jokeString*, and the recipients list needs to be set to *nil* so that the user can choose who the message is sent to. Type:

```
messageComposer.body = jokeString;
messageComposer.recipients = nil;
```

We then want to choose an animation option for the compose view to appear. Feel free to pick whatever style you want, but my code is as follows:

```
messageComposer.modalTransitionStyle =
UIModalTransitionStyleCoverVertical;
```

Finally, we want to present the composer, so we will use this command like we did before:

```
[self presentViewController:messageComposer
animated:YES completion:nil];
```

Oops… I lied about that "Finally" part, this was the final part for the action section, but there is still more! Before we continue, here is all the code that goes in the action:

```
-(IBAction)message {

    MFMessageComposeViewController
*messageComposer =
[[MFMessageComposeViewController alloc] init];
    messageComposer.body = jokeString;
    messageComposer.recipients = nil;
    messageComposer.modalTransitionStyle =
UIModalTransitionStyleCoverVertical;
    [self presentViewController:messageComposer
animated:YES completion:nil];

}
```

It's time to create a local void function. Yes… those exist too. These functions are actually part of the framework they are related to, so they can only be used by that framework and typically don't require extra code to call it.

Our local function will be the handler for the results of the message composer. It looks as such (note: many local functions can't use or have limited use of Xcode's code autofill):

```
(void)messageComposeViewController:
(MFMessageComposeViewController *)controller
didFinishWithResult:(MessageComposeResult)result
{

}
```

Yet again, the result is stored in *result* so that we can use a similar if statement as the other actions. If *result == MessageComposeResultSent*, then we want to let the user know about their amazing success, so we'll use UIAlertView again and then dismiss the message composition. Here it is:

```
if (result == MessageComposeResultSent) {
        UIAlertView *alert = [[UIAlertView alloc]
initWithTitle:@"Congrats, you know how to send a
text" message:@"Your one step ahead of the
elderly" delegate:self cancelButtonTitle:@"Yep"
otherButtonTitles:nil, nil];
        [alert show];
        [self dismissViewControllerAnimated:YES
completion:nil];
    }
```

That's all for sending a text message. In case you have errors and/or missed something the code as a whole is:

```
-(IBAction)message {

    MFMessageComposeViewController
*messageComposer =
[[MFMessageComposeViewController alloc] init];
    messageComposer.delegate = self;
    messageComposer.body = jokeString;
    messageComposer.recipients = nil;
    messageComposer.modalTransitionStyle =
UIModalTransitionStyleCoverVertical;
    [self presentViewController:messageComposer
animated:YES completion:nil];

}

-(void)messageComposeViewController:
(MFMessageComposeViewController *)controller
didFinishWithResult:(MessageComposeResult)result
{

    if (result == MessageComposeResultSent) {
```

```
        UIAlertView *alert = [[UIAlertView alloc]
initWithTitle:@"Congrats, you know how to send a
text" message:@"Your one step ahead of the
elderly" delegate:self cancelButtonTitle:@"Yep"
otherButtonTitles:nil, nil];
        [alert show];

    }
    [self dismissViewControllerAnimated:YES
completion:nil];
}
```

Before we get to Interface Builder, I'd recommend building, but not running, the app by clicking *Product » Build* or by simultaneously pressing ⌘B. If you have any errors check the above code.

6.5 Interface Building Once Again

We're back to our favorite section, the last! Before we start this, I'd like to point out that you cannot use the iOS Simulator to test the text message function, as there is no SIM card or proof of ownership for the simulator to run off of, so DO NOT TRY IT on a simulator or you'll get stuck between a rock and a hard place (literally, I'll come over to your living quarters and smush you against the closest rock)!

Head over to your storyboard. You're going to want to add a "Text View" and four "Buttons" to the view. After that, rename them accordingly and delete all of the UITextView's placeholder gibberish (actually Latin). Then, go to the section titled "View Controller" and then the right arrow button. Once there, do like children (old people term) and connect the dots. Connect each "Received Action" to it's corresponding button (notice I give you less screenshot help).

Now the app is ready to build and run! Remember, DO NOT try to use the message action through the iOS Simulator or really horrible things will happen!

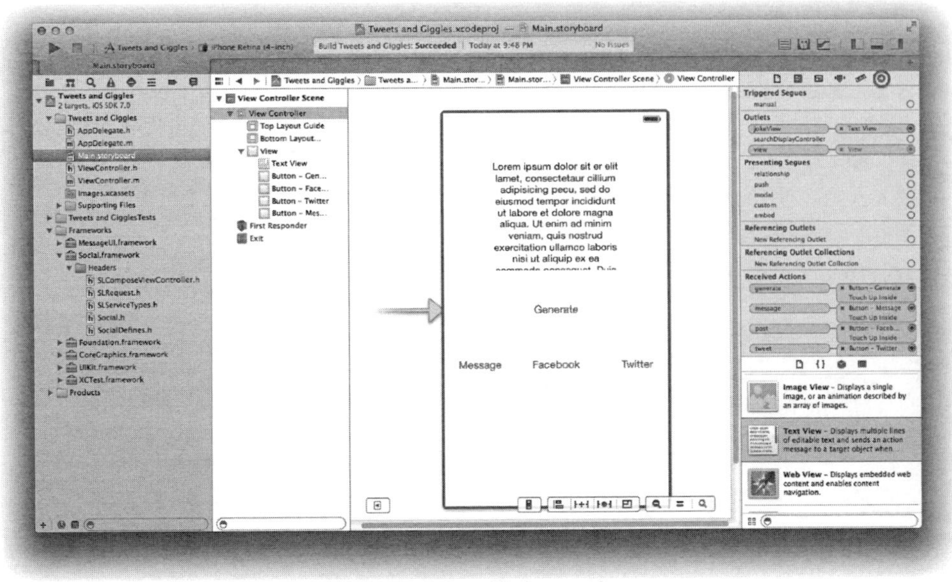

Figure 6.3 - Joke project

Optional: Emailing Some Spam

For bonus points, you must set up *MFMailComposeViewController* in the header file, I'll give you the method file code, but the contents of the header file is all up to you!

The only hint I will give you is to listen to what your errors tell you, they are only there to help (and annoy) you. Very rarely does code run flawlessly the first time around. The method file code is as follows:

```
-(IBAction)email {

    MFMailComposeViewController *mailComposer =
[[MFMailComposeViewController alloc] init];
    mailComposer.delegate = self;
    [mailComposer setToRecipients:nil];
    [mailComposer setSubject:@""];
    [mailComposer setMessageBody:jokeString
isHTML:NO];
    [self presentViewController:mailComposer
animated:YES completion:nil];
```

```
}

-(void)mailComposeController:
(MFMailComposeViewController *)controller
didFinishWithResult:(MFMailComposeResult)result
error:(NSError *)error {

    if (result == MFMailComposeResultSent) {
        UIAlertView *alert = [[UIAlertView alloc]
initWithTitle:@"Yay!" message:@"You emailed
stuff!" delegate:self cancelButtonTitle:@"Doge"
otherButtonTitles:nil, nil];
        [alert show];
    }
    [self dismissViewControllerAnimated:YES
completion:nil];
}
```

Figure 6.4 - The app running

Ch. 6 Conclusion

Another chapter has come and gone, another lesson learned. I actually hope that you've had some errors to learn from along the way! As you may have realized by now, app development, let alone programming, is not the easiest thing in the world to do, but I strongly believe that anyone with enough motivation can learn! Keep persevering though the book, and make sure that at least some of the code makes sense to you. I have to warn you that the next chapter is a toughie, so please make sure you are ready to continue via the chapterly checklist.

- You understand strings well ☐
- You understand what frameworks provide ☐
- You have a decent understanding of functions ☐
- You feel comfortable in the header file ☐

Chapter 7

Mapping the World (At Least Part of It)

Using MapKit to Map the World

Whether you like it or not, Apple Maps has been shoved down our throats and Google Maps has been kicked out the door, but that can't stop (and we won't stop) us from implementing it in our apps. That said, this chapter will have two parts, one focusing on implementing Apple Maps, and the other focusing on implementing Google Maps! I feel that it's important that the user, not the developer, has the option to choose what mapping software they want to use, so you're going to want make sure you follow both tutorials. What we'll be using:

- MapKit framework
- GoogleMaps framework (to be downloaded)
- Many various frameworks and libraries (for Google Maps)
- NSObject files (to hold map pins)
- Latitude and longitude coordinates

To find a pair of latitude and longitude coordinates from a street address, I'd recommend using the website:

`http://stevemorse.org/jcal/latlon.php`

It's a rather barebones website, but gets the job done well.

7.1 Setting Up the Apple Maps Project

Just like our other projects, we're going to create a new Xcode project that's based under the "Single View" template. Name the project as you wish and save it somewhere that you'll remember where it is in a time of need. I always say this because there will probably be times in your future where you want to implement something in an app and need to go back to a project you previously made to find the code because, you know, it's hard to remember stuff (especially code). Create the project and continue.

Ah… a blank slate again! Since Apple Maps is the default mapping program, the implementation is rather simple (compared to Google Maps!). All we'll need framework wise for Apple Maps is the MapKit framework. To import the MapKit framework go to *Mapping the World » Target: Mapping the World » Build Phases » Link Binary With Libraries »* + (note: "Mapping the World" is a variable for displaying your project's name) and import *MapKit.framework.*

Figure 7.1 - Adding frameworks

Now that the frameworks have been added, we want to make sure that our ViewController gets to reap the benefits of MapKit, so head over to the header file and add this to the top:

```
#import <MapKit/MapKit.h>
```

Once added, the Apple Maps libraries are included into the build phases which allow for use in our code.

The only other things that we need to add to include a view for our map and a couple buttons that toggle between the different options of presentation (hybrid, satellite, and standard). The map view will be an outlet, and the toggle buttons will be action functions. Let's jump into the code (don't forget to make the header file into two sections):

```
IBOutlet MKMapView *mapView;
```

In the function section, add the actions:

```
-(IBAction)mapToHybrid;
-(IBAction)mapToSatellite;
-(IBAction)mapToStandard;
```

We're ready to move onto the method file, but first, check your code with mine:

```
#import <UIKit/UIKit.h>
#import <MapKit/MapKit.h>

@interface ViewController : UIViewController {
    IBOutlet MKMapView *mapView;

}

-(IBAction)mapToStandard;
-(IBAction)mapToHybrid;
-(IBAction)mapToSatellite;
@end
```

Starting to Get Down and Dirty

Now that our map is set up (partially… hehe), we can douse the project with code! Head over to the method file to get going.

To do some preliminary work, add the action functions we made in the header to the method file as such:

```
-(IBAction)mapToHybrid {

}

-(IBAction)mapToSatellite {

}

-(IBAction)mapToStandard {

}
```

Now we need to go back to *viewDidLoad* to set up part of the map view. We should set a default type of map to be presented upon launch. I personally prefer the "hybrid" option, but many prefer the "standard" option, and the "satellite" only option is also available. Some other things that are handy for the user is the ability to scroll and zoom… imagine trying to find a street when the map is zoomed all of the way out! A little side note, the scroll option and zoom option are boolean statements, if you remember boolean you'd know that there are only two options, yes or no.

> "Do. Or do not. There is no try."
>
> - Yoda (894 BBY - 4 ABY)

The code looks as such:

```
- (void)viewDidLoad {

    mapView.mapType = MKMapTypeHybrid;
    mapView.scrollEnabled = YES;
```

```
    mapView.zoomEnabled = YES;

    [super viewDidLoad];
}
```

As you may have predicted, the same line of code that sets the type of map will be used inside of our action functions in order change the type of map whenever the corresponding button is tapped. Go ahead and try to add this code on your own and then continue. This is what the actions should look like:

```
-(IBAction)mapToHybrid {

    mapView.mapType = MKMapTypeHybrid;

}

-(IBAction)mapToSatellite {

    mapView.mapType = MKMapTypeSatellite;

}

-(IBAction)mapToStandard {

    mapView.mapType = MKMapTypeStandard;

}
```

Optional: Once this is completed, you can now add a "Map View" and four "Buttons" to your Interface Builder view, connect them to their corresponding actions and outlets, and run the app to see if it works from here. I will not be covering this right now with detailed step-by-step instructions, but I will once the project is finished. The purpose of you doing it by yourself is to challenge you as best as I can. If everything doesn't go right, just continue on for now and follow my step-by-step interactions later.

7.2 Creating, Importing, and Connecting Object Files

Due to some current limitations with the MapKit framework, in order to use annotations (or pins), a separate set of files assigned as objects and later, annotations, must be created. Don't worry though, it's not too difficult. In order to create an object file (or any file) click through *File » New » File...* and you'll be taken to a screen that asks what kind of file you want to create. In the left-hand side, make sure that "Cocoa Touch" is highlighted. From there, select "Objective-C class" and then click "Next" to be taken to the further selections page.

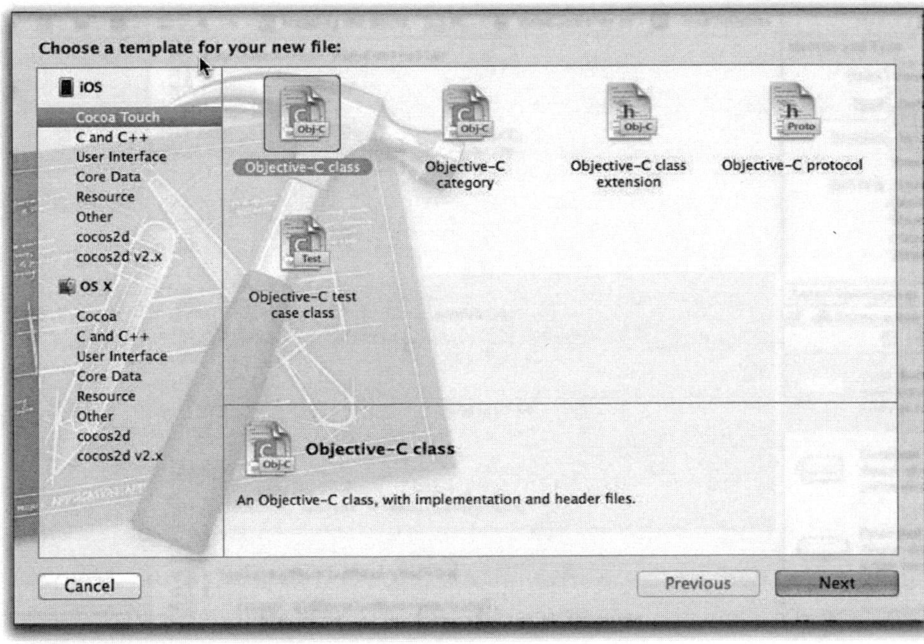

Figure 7.2 - File template selection

On this options page, you'll have to name the file and tell Xcode what kind of file subclass you want the file to be assigned as. Under the “Class” field, type whatever you want, as this field assigns the name of the header and method files that will be created. For the “Subclass of” field, select the option “NSObject” as the subclass. Once finished click “Next”.

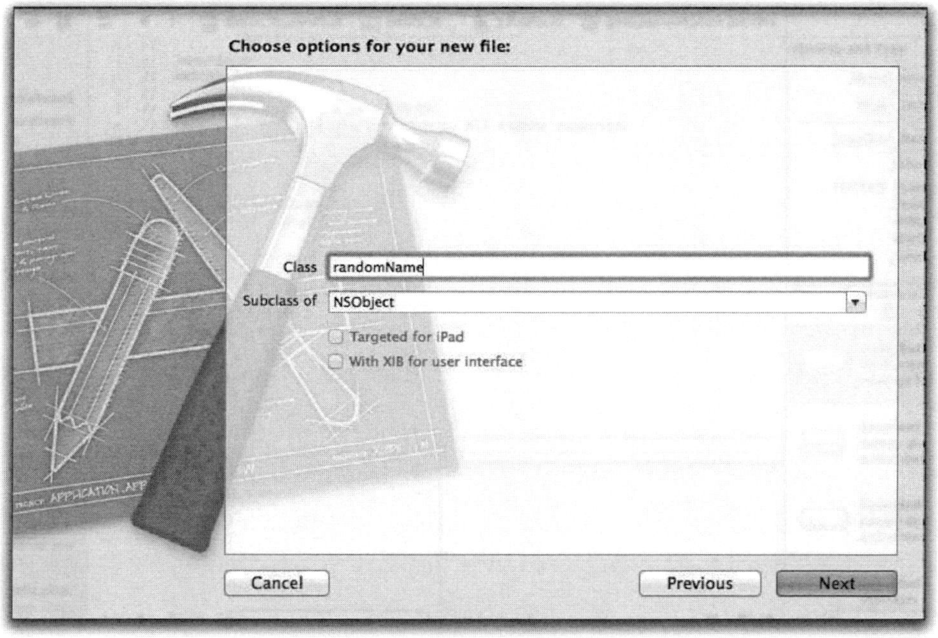

Figure 7.3 - File creation options

The final page will ask you where you want to save the file, DO NOT change the default save location, click “Finish” and be on your jolly way.

Head over to the newly created files’ header file. The first thing we need to do is import the MapKit framework, although, it’ll be different this time. In order to save space, we only want to import the part of MapKit that handles the pins. This means we must change the typical */MapKit.h* to the file we want, which is */MKAnnotation.h*:

```
#import <MapKit/MKAnnotation.h>
```

After importing, we want to change some of the object settings by even further describing the file. We want to specify it as an annotation and only an annotation. To do this we want to add

<MKAnnotation> next to the line *@interface fileName: NSObject* and before the opening curly bracket (don't forget to add the curly brackets if you haven't already). That line should now look like this:

```
@interface fileName : NSObject <MKAnnotation>
```

Okay, now it's going to get tricky. In the first section of your header file add the following:

```
CLLocationCoordinate2D coordinate;
NSString *title;
```

CLLocationCoordinate2D is an object that will specify the coordinates that we want the pin to be located at. The string will define what the pin will say whenever tapped.

Okay, move onto the next section of header file code (after the closing curly bracket). Here we are going to want to turn the variables that we just defined into properties. In Objective-C, properties can be variables or functions that are going to be used in an unrelated file (such as the relationship between our new file and the default files). Because we want to use this object annotation file in our main view controller, we must "properatize" our variables. The variables as properties look like this:

```
@property (nonatomic, assign)
CLLocationCoordinate2D coordinate;
@property (nonatomic, copy) NSString *title;
```

You may have noticed *nonatomic, assign*, and *nonatomic, copy*. These are instructions for the program to follow when making the new property. The code *nonatomic* means that that property can be used in other files, as opposed to *atomic*, which doesn't allow sharing. The *assign* code is simply a default that must be placed there whenever nothing else needs to be said (kinda like *nil*). The code *copy* copies the current values of the property and saves it for later use. This is used to prevent data corruption among different files sharing information.

Now head over to this file's corresponding method file so we can do something called synthesization. Synthesizing a file helps the method file keep track of the property, and also helps manage the memory levels. Synthesizing is required of all properties, so it's something you just gotta do even though it doesn't do much for you:

```
@synthesize coordinate, title;
```

Add that piece of code anywhere between *@implementation* and *@end*. It's already time to move back to our main header file!

Once at the original header file, import our new files' header file into the original like this:

```
#import “fileName.h"
```

Please note that we used " " instead of < >. This is because you only use < > for importing frameworks, and use " " for importing everything else. It's already time to go to the original method file!

Getting Down and Dirty

In this section, we'll be defining the coordinates of our pin, so if you haven't already, type in an address on: http://stevemorse.org/jcal/latlon.php to get the matching longitude and latitude.

Got a pair of coordinates? Great! Save them somewhere and we can continue on to coding! Okay, find the spot in *viewDidLoad* where we set the properties of *mapView*. Found it? Add a space under it (still in *viewDidLoad*), as this will be where we'll add the coordinate code.

In order for an annotation to know where to locate itself, we need coordinate regions. Let's start coding some regions! First off, we need to locally declare a coordinate region. Regions want to know the details as soon as possible, so they require some preliminary numbers. Since these numbers aren't really needed, they'll be set to zero:

```
MKCoordinateRegion region = {{0.0, 0.0}, {0.0,
0.0}};
```

MKCoordinateRegion variables just happen to be irregular, and don't require an asterisk (*) before setting the name. The next step is setting the latitude and longitude. In a literal sense, you will be "centering" the code on the latitude and longitude:

```
region.center.latitude = 37.331741;
region.center.longitude = -122.030744;
```

The latitude and longitude can be as precise as you want, which all depends on the amount of numbers you have defining the location. A good rule of thumb is to use at least six numbers following the decimal point. Next up we have the amount zoomed in (or "spanned") by the program default. This is sort of confusing, and goes as such:

```
region.span.longitudeDelta = 0.003f;
region.span.latitudeDelta = 0.003f;
```

The closer your number is to zero, the more zoomed in the map will be by default. The trailing *f* is to represent that it is a floating point number (remember float!), which in this use means that the number doesn't have to be exact, and that the computer will choose the best number nearest to the set number. To finish up with the region, we need to set our map's area of focus onto the region just defined:

```
[mapView setRegion:region animated:YES];
```

Just to recap, we just created and set a coordinate region equal to the coordinates found online, and then set our map's focus area on these coordinates:

```
MKCoordinateRegion region = {{0.0, 0.0}, {0.0,
0.0}};
    region.center.latitude = 41.242142;
    region.center.longitude = -80.448654;
```

```
    region.span.longitudeDelta = 0.003f;
    region.span.latitudeDelta = 0.003f;
    [mapView setRegion:region animated:YES];
```

"Okay… all of this is interesting, but why exactly did we create new files, and import them into our main view?" Good question self! You're about to find out just why.

Add another space down from the code we just created, as we're about to finally use our object file! In order to interact with the object file, we must declare it as if it is a local variable (*pin* is my object file's name):

```
pin *annotation = [[pin alloc] init];
```

Now, the *annotation* variable is now sort of a delegate for the object files, and this main file and the object files are connected in a way. If you want to access anything declared in the object files, you simply add *annotation* before it! We now want to give *annotation*'s variables something to write home about (as if variables had homes or someone to write to):

```
annotation.title = @"My annotation!";
annotation.coordinate = region.center;
```

The *annotation* variable is now all set and good! As of right now through, *mapView* isn't planning on showing off any annotations. In order to fix this we need to tell it to add *annotation*:

```
[mapView addAnnotation:annotation];
```

Yay! We're all good now with coding an Apple Map and adding a pin to it at a custom location! Next up is connecting everything in Interface Builder (if you didn't already do it yourself). In case you have any errors, or are confused, here is the finished method file code:

```
- (void)viewDidLoad
{
```

```
    mapView.mapType = MKMapTypeHybrid;
    mapView.scrollEnabled = YES;
    mapView.zoomEnabled = YES;

    MKCoordinateRegion region = {{0.0, 0.0},
{0.0, 0.0}};
    region.center.latitude = 41.242142;
    region.center.longitude = -80.448654;
    region.span.longitudeDelta = 0.003f;
    region.span.latitudeDelta = 0.003f;
    [mapView setRegion:region animated:YES];

    pin *annotation = [[pin alloc] init];
    annotation.title = @"My annotation!";
    annotation.coordinate = region.center;
    [mapView addAnnotation:annotation];
    [super viewDidLoad];
}

-(IBAction)mapToHybrid {

    mapView.mapType = MKMapTypeHybrid;

}

-(IBAction)mapToSatellite {

    mapView.mapType = MKMapTypeSatellite;

}

-(IBAction)mapToStandard {

     mapView.mapType = MKMapTypeStandard;

}
```

Interface Building

Open up the storyboard file to get started. In here, you're going to want to add one "Map View" and three "Buttons" from the objects list to the view. Arrange them how you please. Click "View Controller" and then the right arrow from the top-right corner. Then, connect your *mapView* to the actual "Map View", and the three action functions to their corresponding buttons.

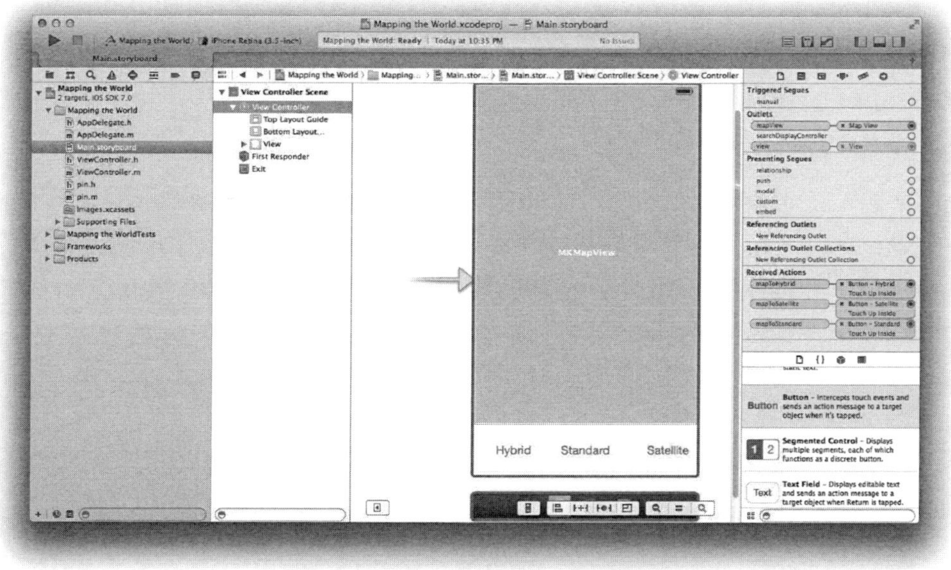

Figure 7.4 - Interface building

Once all connected, run the app through the method of your choice and you should be presented with a map that focuses on the region whose coordinates you typed in. At launch, the app will ask if it can use your location, please note that in iOS Simulator, the location is always defaulted to the Apple Headquarters at 1 Infinite Loop, Cupertino, CA. You can change this location via *Debug » Location...* in the simulator.

We're ready to move onto Google Maps, your final non-optional project! Don't fret though, there are two more chapters, but they focus on specific things that are good to know, but not completely necessary.

7.3 Google Maps (Importing SDKs and Libraries)

Whoa, we've come so far! Our step-by-step project journey is almost over, but there is definitely at least one more thing that every developer needs to know. This is how to use SDKs (software development kits).

SDKs are vital to seamlessly adding creative designs and expanding the possibilities of your app. SDKs are typically sponsored by a company that want you to implement their services into your app. A great place to find SDKs and other components is at: www.cocoacontrols.com. For example, the Google Maps SDK will allow you to implement Google Maps. Speaking of Google Maps, let's get into this!

So we want to implement Google Maps, but ever since Apple Maps, Google Maps doesn't come included with Xcode. I bet a Google Maps SDK for iOS (what a long shot from the developers of Android) would help… oh wait! There is a Google Maps SDK for iOS and it can be found at:

`https://developers.google.com/maps/documentation/ios/`

Upon arriving on the page, click "Download the SDK" to be taken to yet another page. On the next screen, click the highlighted version number button to download the official SDK.

Figure 7.5 - Downloading the Google Maps SDK

The download (at the time of writing) is about 18MB, so it shouldn't really take too long. Once downloaded, place it somewhere memorable as we'll come back to it later. Leave this page open because we're going to use it later to register for an API key.

Setting Up the Google Maps Project

First of all, we're going to want to create a new "Single View Application" project. Name and save it as you please.

Now that we have a blank slate again, we're going to have to set some things up. In order to use Google Maps, we must import the Google Maps framework. Head over to *ProjectName » Build Phases » Link Binary With Libraries »* +. Now add the Google Maps framework… oh yeah, Xcode no longer includes it. So how do we add a third-party framework? It's pretty simple, next to the "Cancel"

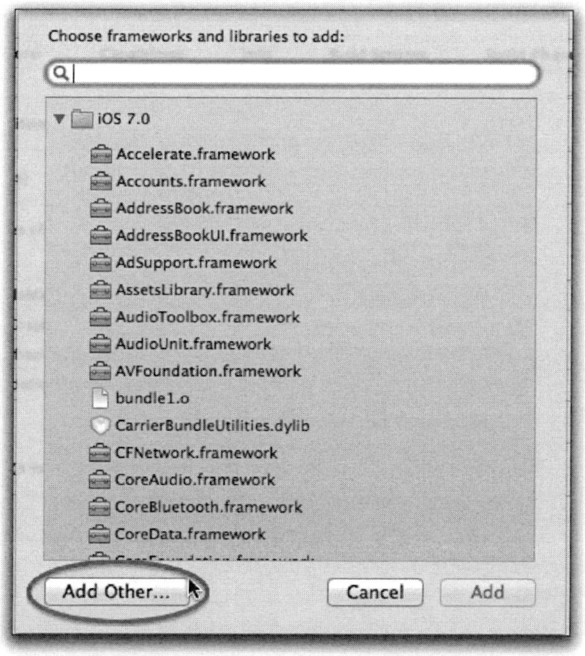

Figure 7.6 - Adding third-party frameworks

and "Add" button there is a button with the text "Add Other…". Click "Add Other…", and you'll be presented with the standard file picker. Find the framework, *GoogleMaps.framework* located in the SDK folder that you downloaded just a little bit ago (you may still have to unzip the SDK folder first if not done so already). Add the framework and notice that it has been added to "Link Binary With Libraries", and looks just like the other frameworks. Don't exit this section just yet as there is still more to add.

Yet another downfall of Google Maps no longer being integrated with Xcode is that all of the necessary frameworks and libraries are not automatically imported when *GoogleMaps.framework* is imported. Sadly, there are a whole bunch of frameworks that need linked to the binary. Here's the list of dynamic libraries (.dylib) and frameworks that need to be linked (for the .dylib files, I'd recommend using the search function to find them):

- AVFoundation.framework
- CoreData.framework
- CoreLocation.framework
- CoreText.framework
- GLKit.framework
- ImageIO.framework
- OpenGLES.framework
- QuartzCore.framework
- SystemConfiguration.framework
- libc++.dylib
- libicucore.dylib
- libz.dylib

Don't worry, we aren't going to have to import and use these frameworks and libraries in our code, but they're used by the Google Maps framework, so we need to import them in order for it to work properly.

Now, we need to import the bundle file (.bundle) for Google Maps which is located inside of *GoogleMaps.framework*. I truly don't know why it is required to be imported whenever it is inside of a file already imported, but it is, so... find the framework in the location it was downloaded to, open it (double click), open the folder named, "Resources", and drag *GoogleMaps.bundle* into your Xcode project. Whenever asked about copying the file, make sure that you DON'T copy the bundle.

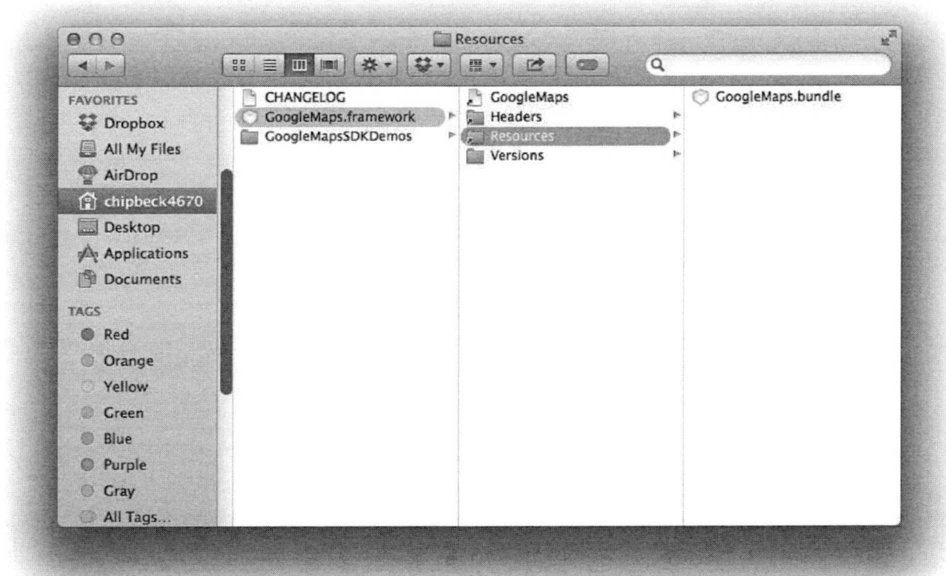

Figure 7.7 - Adding a bundle

The last preliminary thing we need to do is to add a linker flag. A linker flag allows for all code to be packaged with the app that is connected to the flag. To get started go to the "Build Settings" tab of the target screen. From there scroll through *Linking » Other Linker Flags*.

Double click the blank space on "Other Linker Flags" and type in " -ObjC ". Press enter and you're done!

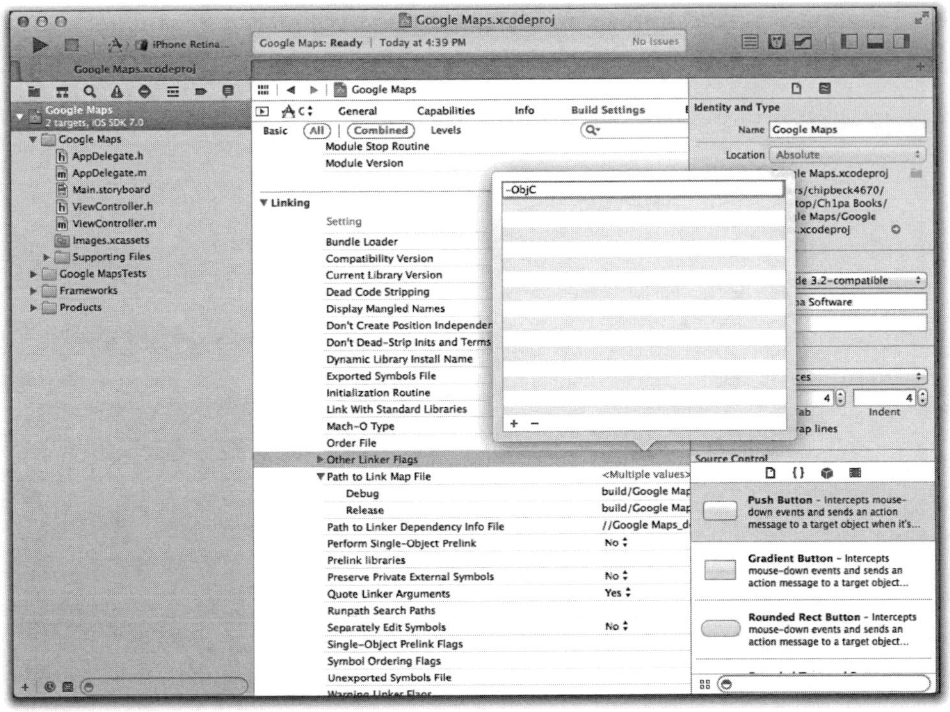

Figure 7.8 - Linker flagging

Registering an API Key

Okay, I hoped you left that webpage open where you downloaded the SDK because we still need that page! If you closed it, simply reopen the page because there was nothing we entered before that is vital to this step. Locate the section titled "Obtaining an API Key" and click on the link that says “Google APIs Console”. From there, click on the “Services” tab towards the right side of the screen. Scroll

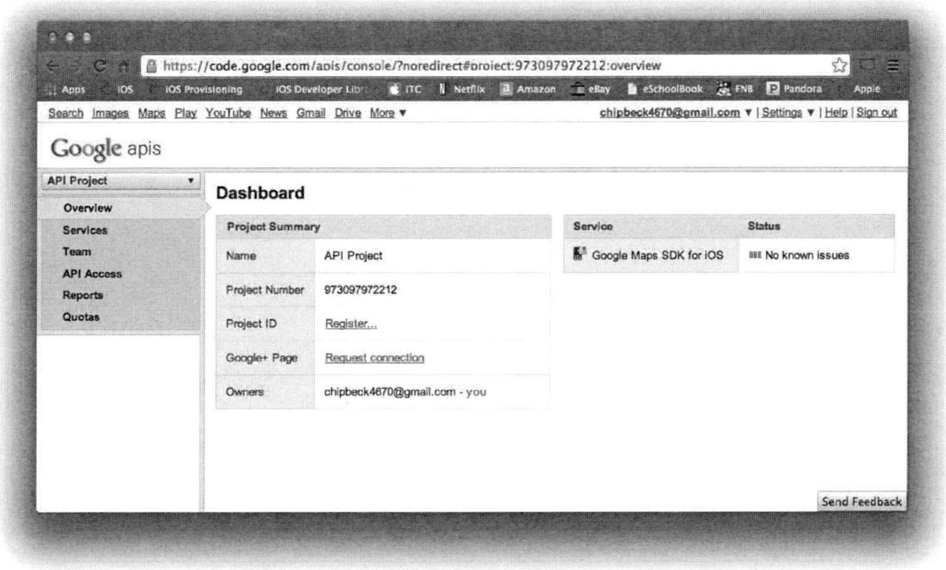

Figure 7.9 - Google APIs dashboard

down to find “Google Maps SDK for iOS” and slide the switch to turn it on. You’ll then have to “agree” to the Terms of Service. Once completed, click on the “API Access” tab. Click on the “Create new iOS key…” button to be presented with a screen that asks for the bundle identifier of the app that will be using the API key. Type in

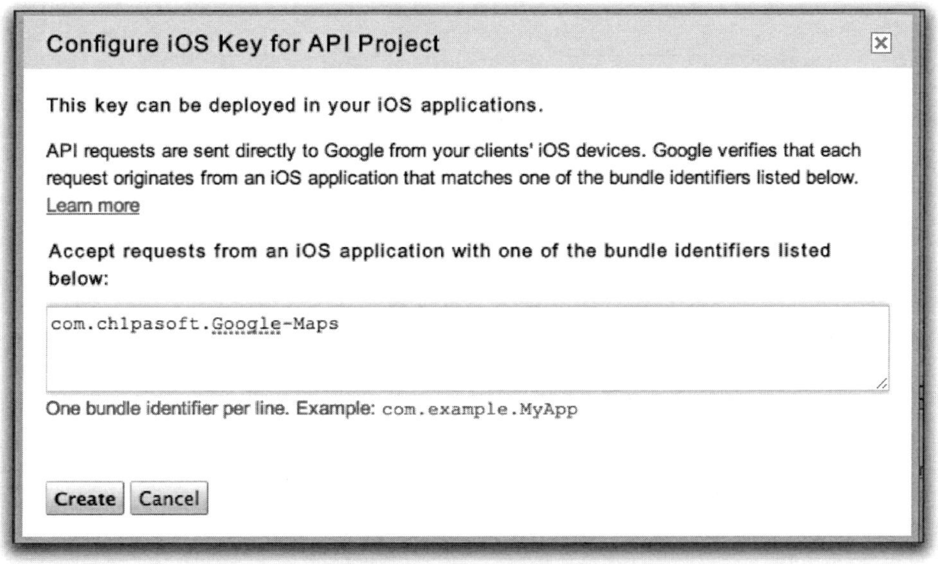

Figure 7.10 - Configuring iOS API key

the bundle identifier, and then press "Create" for Google to generate a key for your identifier.

Once Google is done generating your key, you'll find it in the "API Access" section. Copy the key, and save it for later.

Setting Up the Google Maps Project Pt. 2

Whew! I bet you're glad to know that we're done with configuring app information online, and we can get to coding!

Go back to Xcode, and go to the file *AppDelegate.m*. No, we've never worked with the *AppDelegate* file before, but we're about to now.

The *AppDelegate* is like an extreme version of *viewDidLoad*. The major difference is that *AppDelegate* handles events like termination, going into the background (multitasking), returning from the background, and what to do upon launch.

The first step is to import the Google Maps framework:

```
#import <GoogleMaps/GoogleMaps.h>
```

Then, in the first section, *- (BOOL)application:(UIApplication *)application didFinishLaunchingWithOptions:(NSDictionary *)launchOptions*, we're going to register the app with the API key just generated. The prefix for the Google Maps framework is "GMS". The code is as follows:

```
[GMSServices provideAPIKey:@"YOUR_KEY"];
```

That's it for *AppDelegate.m*! Time to move onto the view controller header file.

In the header file, we must first import Google Maps, then separate the header file into two sections as we have done many times before. Next we'll declare a Google Maps map view and then create four actions for changing the type of map displayed (yes there are four, standard, satellite, hybrid, AND terrain). The header file will then look like this:

```
#import <UIKit/UIKit.h>
#import <GoogleMaps/GoogleMaps.h>

@interface ViewController : UIViewController {

    GMSMapView *mapView;

}

-(IBAction)toSatellite;
-(IBAction)toStandard;
-(IBAction)toHybrid;
-(IBAction)toTerrain;

@end
```

Now let's move onto the method file. Now it gets… easy! All this other crap we had to do earlier pays off now because the implementation is quite easy.

The Meat of Google Maps

We first have to define the camera position which will act as both the default center of the map, and the spot for the marker (Google's take on pins/annotations). We'll reuse the latitude and longitude of the previous project. The last camera property we need to define is the zoom. Google Maps is odd in the fact that the minimum is zero, while the maximum varies the max, depending on location, can be thirty, forty-five, or sixty-five and a half. Let's look at the code:

```
GMSCameraPosition *camera = [GMSCameraPosition
cameraWithLatitude:41.242142 longitude:-80.448654
zoom:15];
```

Next, we'll declare where the map is to be located, and what it's size should be. Since there is no Google Maps template in Interface Builder, we have to make it appear programmatically! This is done through coordinates, the first two coordinates determine the location, with zero, zero being the top-left corner. The third and forth coordinates represent the size, width by height. We don't want to cover the whole screen as there needs to be room for the buttons, so we'll only cover about three-fourths of the screen. The last property needed is the camera. Luckily we just declared a camera. We'll set the map view like this:

```
mapView = [GMSMapView mapWithFrame:CGRectMake(0,
0, 320, 400) camera:camera];
```

After this, we need to create a marker at this location. First, allocate some memory, then give it a position, and finally give it a title and subtitle (snippet = subtitle):

```
GMSMarker *marker = [[GMSMarker alloc] init];
    marker.position = camera.target;
    marker.title = @"Google Maps";
    marker.snippet = @"iOS App Development for
Teens by Teens";
```

Finally we need to set the marker on the map, and set the map on the view:

```
[marker setMap:mapView];
[self.view addSubview:mapView];
```

That's it for setting up the map; it's a lot simpler than adding pins to Apple Maps! Here's the code so far as a whole:

```
- (void)viewDidLoad
{

    GMSCameraPosition *camera =
[GMSCameraPosition cameraWithLatitude:41.242142
longitude:-80.448654 zoom:50];
```

```
    mapView = [GMSMapView
mapWithFrame:CGRectMake(0, 0, 320, 400)
camera:camera];
    GMSMarker *marker = [[GMSMarker alloc] init];
    marker.position = camera.target;
    marker.title = @"Google Maps";
    marker.snippet = @"iOS App Development for
Teens by Teens";
    [marker setMap:mapView];

    [self.view addSubview:mapView];

    [super viewDidLoad];

}
```

Now let's create the actions to change the map type (don't worry it's easy). I'm just going to hand it out because it probably insults your intelligence:

```
-(IBAction)toHybrid {

    mapView.mapType = kGMSTypeHybrid;

}

-(IBAction)toSatellite {

    mapView.mapType = kGMSTypeSatellite;

}

-(IBAction)toStandard {

    mapView.mapType = kGMSTypeNormal;

}

-(IBAction)toTerrain {
    mapView.mapType = kGMSTypeTerrain;
}
```

Doing a Little Interface Building

There's not much to do in Interface Builder since the map is added programmatically. Open the storyboard, and add four buttons to the view. All you gotta do now is connect the "Received Actions" to their corresponding buttons and you'll be all set!

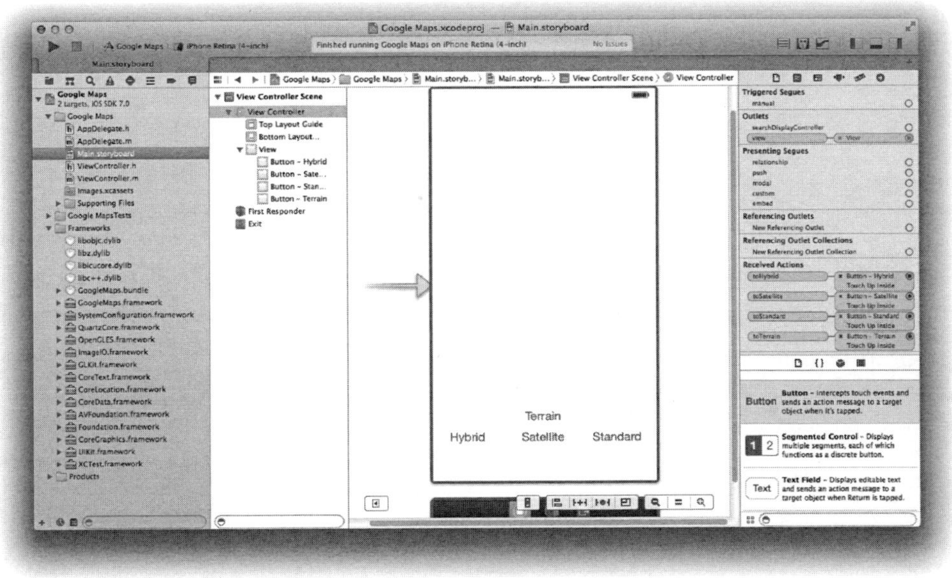

Figure 7.11 - Google Maps Interface Building

You're now ready to build on your selected device!

CH. 7 CONCLUSION

That was one big fish to fry, but there are tons of fish in the sea (I know that this is not used in the right context, but who cares!) and you have just caught a little guppy (contradicts my big fish theory (see what I did there (big bang theory))). You've just scratched the surface. There are limitless possibilities to developing apps! You are now an Apple Maps and Google Maps connoisseur! In case you choose not

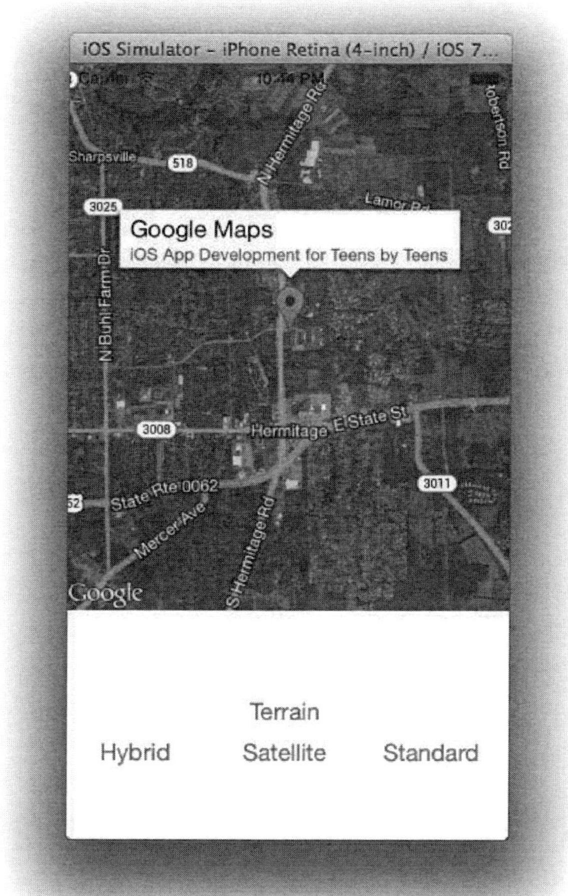

Figure 7.12 - Google Maps in the simulator

to read the last two chapters, I thank you for your time with me and wish you the best of luck on your future endeavors!

- You can build an app on your own by now ☐
- You can successfully import frameworks and libraries ☐
- You know where to get coordinates ☐
- You can connect buttons to actions ☐

Chapter 8

What's Important in Life

Some Very Useful Things to Know How to Do

I hope you don't think you're done learning! We still have two more chapters to cover and a whole bunch of important topics! Here you'll learn everything that I felt should be included, but are too small for their own chapter and wouldn't mix well in the already existing chapters.

In this chapter, you'll learn about the following:

- Adapting between the iPhone and iPad
- Working with other apps through URL schemes
- Instagram posting
- Opening/previewing files
- Math in programming
- App icons
- App display name

Don't worry if you don't understand the items covered in this section, most will come over time, and the rest will come to you by rereading it (life experience)! Some of these items are more complicated than the others, but that doesn't mean you still can't have fun! If things get too complicated, just go back to the earlier chapters, and read on.

Before starting this section, please note that it will be minimalistic. I will not give many setup instructions, I want you be able to create a project all by yourself. Every section in this chapter can be completed using a "Single View Application" template (create a new project every section).

8.1 Adding iPhone and/or iPad Specific Code

Problem: You have separate iPhone and iPad views that share the same view controller, but some code is specific to the iPhone and other code is specific to the iPad.

Solution: Have a boolean statement define whether the device is an iPhone or iPad through an if statement in *viewDidLoad.*

So you want to be able to define what code runs on which device. This is solved through the simple use of a boolean statement whenever the view loads. Let's get started!

When setting up and creating the project, make sure you select the "Universal" option so that two storyboards for the iPhone and iPad are created.

Once created, head to the header file (haha) and divide it into two sections as usual. In the first section, we're going to want to initialize a boolean statement as *BOOL*. There are many other options similar to *BOOL*, but make sure you type it as shown above. *BOOL* is another one of those irregular variables in Objective-C that doesn't use an asterisk (you still can, but you'll get an error). Our statement looks as such:

```
@interface ViewController : UIViewController {
 BOOL isiPad;
```

```
}
@end
```

Pretty simple right?! Okay, head over to the method file so that we can polish this baby up.

Let's suppose that the code that needs to be different is for the iPad (it typically is). In order for this to be true, the width of the screen must be greater than three-hundred twenty, or the width of the iPhone. But how do you find the screen size? This is through using *UIScreen* and getting deep into it by calling *mainScreen*, or the current screen, and then *bounds*, or dimensions, and even further by *size.width*. Place this inside of *viewDidLoad*:

```
if ([[UIScreen mainScreen] bounds].size.width >
320) {

        isiPad = YES;

    }
    else {

        isiPad = NO;

    }
```

Not bad, not bad at all! Right now, if the current dimensions exceed three-hundred twenty then the boolean statement, *isiPad*, will set it's value to *YES* (which can be translated to *TRUE* or *1*).

Now for the implementation. We want to write a simple if statement that runs if *isiPad* is equal to *YES* (remember that == is used to compare, and = is used to set a value):

```
if (isiPad == YES) {
//code here
}
```

8.2 URL Scheming

Problem: You want to communicate to an existing app through your app, make a call, open a webpage, or open up pretty much any other app.

Solution: Test the device's ability to open the app through URL schemes, and then use a system command to open the scheme. (Websites for find URL schemes: http://wiki.akosma.com/IPhone_URL_Schemes, http://handleopenurl.com/scheme)

There are these great things called URL schemes that every app has once created (although assigned randomly by default) that allows the app to be accessed by typing a URL. A major URL scheme I know you use daily is "http", or hyper text transfer protocol! Typing "http" allows the device to know where to go to send the information following it. Every URL scheme is followed by a colon and two forward slashes (for example: http://), then further instructions follow (http://www.blahblah.com). As for apps, just typing "name://"opens the app that is registered with that scheme. Let's get started by creating a new project.

Once you have a project created, go to the header file. Divide it into two sections as usual. Once divided, declare a URL in the first section:

```
NSURL *url;
```

In the next section of the header, initialize three actions, one for opening a webpage, one for opening your Facebook newsfeed, and one for making a phone call:

```
-(IBAction)openWebPage;
-(IBAction)openFacebook;
-(IBAction)makeCall;
```

Now we can move over to the method file. Skip over *viewDidLoad* and declare the actions that we just initialized.

Opening a Link in Safari

Inside of *openWebPage*, we're going to assign *url* to a standard website URL:

```
url = [NSURL URLWithString:@"http://
www.Ch1paSoft.com"];
```

Since every Apple device (including iOS Simulator) can open a webpage, we don't need to place anything to check if the device can open that URL (we will later). Let's finish this action off by opening the URL by accessing the deepest part of the application :

```
[[UIApplication sharedApplication] openURL:url];
```

Facebook App URL Schemes

Now move to the Facebook app newsfeed opening action. Since every person that owns an iOS device doesn't have the Facebook app, we have to assign *url* and then check to see if our device can open it:

```
url = [NSURL URLWithString:@"fb://feed"];
    if ([[UIApplication sharedApplication]
canOpenURL:url]) {

    }
```

Inside of the if statement, we are going to put the command that opens the Facebook newsfeed:

```
[[UIApplication sharedApplication] openURL:url];
```

You may have noticed that I specifically mentioned "newsfeed" in this section. That's because there are tons of other things that you can open in the Facebook app via URL schemes. Here is a list of the schemes that you might most commonly use:

```
fb://profile
```

```
fb://friends
fb://notifications
fb://feed
fb://events
fb://requests
fb://notes
fb://albums
fb://birthdays
fb://groups
fb://mailbox
fb://map
fb://messaging
fb://upload
```

There are still tons more! You can find more at the link that I mentioned earlier (http://wiki.akosma.com/IPhone_URL_Schemes, http://handleopenurl.com/scheme).

Making Phone Calls

Because Apple doesn't want developers to be able to develop apps that make hidden spam calls, the only way to programmatically make a telephone call (call has multiple meanings) is through a URL scheme.

Head over to the *makeCall* action. We're going to assign *url* to the Phone app while simultaneously dialing a number. Then, we must determine if the device can make phone calls, and open it:

```
url = [NSURL URLWithString:@"tel:
1-800-356-9377"];
    if ([[UIApplication sharedApplication]
canOpenURL:url]) {

        [[UIApplication sharedApplication]
openURL:url];
```

```
}
```

A very odd thing about this URL scheme and the schemes of all of Apple's built-in apps (excluding Safari) is that they do not use two forward slashes after the colon.

"*tel:*" URLs can contain numbers, dashes, and "+"s, but not spaces or brackets.

We're done with the coding for launching other apps, so go to the storyboard, add three buttons, connect them, and run the app to test it (Note: iOS Simulator will only work with the Safari link, the Facebook scheme requires the installation of the Facebook app on a physical device, and the phone scheme only works on the iPhone). Now onto creating our own URL scheme!

Custom URL Schemes

In whatever application you choose, under "Supporting Files", open the file with the extension ".plist". Property list (plist) files typically contain various settings. This is where we will add our app's URL scheme.

Right click anywhere in the property list and click "Add Row". When clicked, you'll be presented with a whole list of options. Either scroll down and find or type "URL types". Open that option to "Item 0", then open that to "URL identifier". Change "URL identifier" to "URL Schemes", and expand it to find "Item 0". Under this "Item 0" change the blank value on the right hand side to whatever URL scheme you would like. Leave out the "://" because it is automatically included, but never shown. For example, “chip” will become “chip://” in the code.

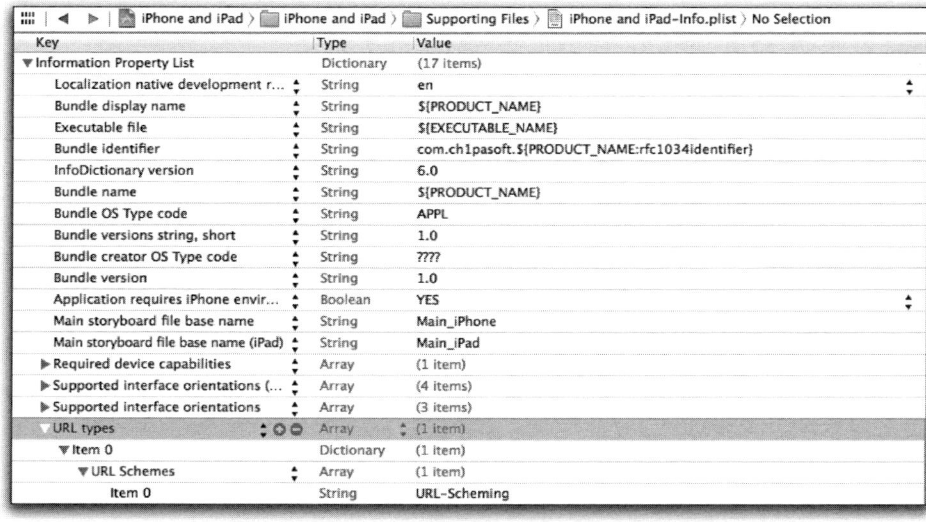

Figure 8.1 - Adding to the property list

8.3 App Display Name

Problem: When you initially created a project you weren’t sure about the name, or you want to change the current name.

Solution: The property list (plist) file discussed in the last section contains the app’s editable display name.

Whenever you gave your project a name during the creation process, that name is given to a whole bunch of other things, the “Bundle identifier” (not editable), “Bundle name” (not editable), and the “Bundle display name” (editable). The value of the “Bundle display name” is read by the device it’s running on, and places it under the app icon. This becomes problematic whenever you want to change the name chosen during creation. Fear not! This can be changed very easily by taking a little trip down to your local property list! Open your projects property list, find the section titled "Bundle display name" and change the default value, "${EXECUTABLE _NAME} to whatever name you want. The field can carry

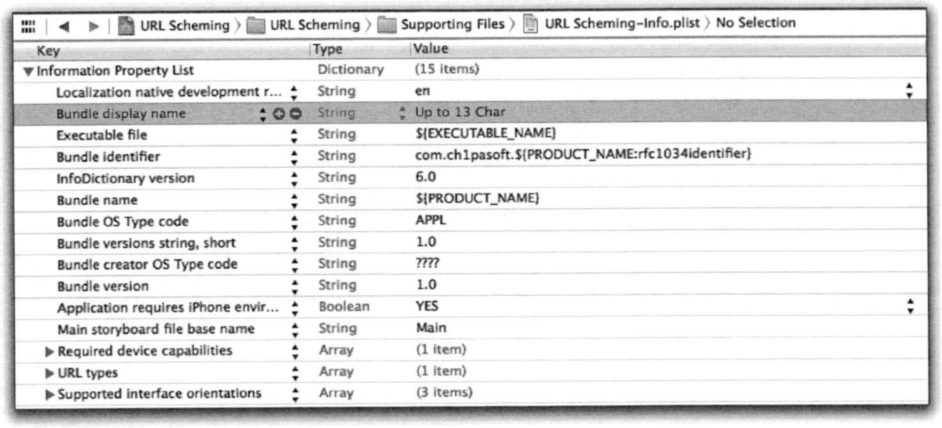

Figure 8.2 - Changing the display name

approximately eleven characters (depending on the size of the character). Whenever it has over thirteen characters, the end is truncated with "…". The name looks very bad with "…" following it, so I'd recommend shortening your name if possible.

8.4 App Icons

"Don't judge an app by it's cover." Bullcrap! There is no way someone will tap on your app in the App Store if you have a crappy icon "just cuz". Every app needs a good (and accurate) icon!

Figure 8.3 - The "Bundle display name"

Adding icons can be a bit tedious due to the fact that you need four icons for the iPhone and seven icons for the iPad! The reason you need so many icons is because you need a proper size for advertising, for the actual icon, and for the supported non-Retina display devices. Here's the rundown

iPhone

- 58x58
- 80x80
- 120x120
- 1024x1024 (iTunes)

iPad

- 29x29
- 40x40
- 58x58
- 76x76
- 80x80
- 152x152
- 1024x1024 (iTunes)

Design your icon with the dimensions being 1024x1024. In order to shrink the big file to fit all of the dimensions above, I recommend using an app like Preppo from the App Store. There are also online resources to do this, but you don't know what they're doing with your file (stealing). Once you have an icon of every dimension, go to the "General" tab of your project's settings, scroll down to "App Icons", click on the little right arrow button to be taken to a screen where you can drag and drop your icons.

On the drag and drop screen, make sure you are under the item "AppIcon". If you are not, please make sure you do so. Once there, as stated before, drag and drop all your icons to your hearts content! If you get any errors or yellow signs, that means that the file you dropped is not of the correct dimensions.

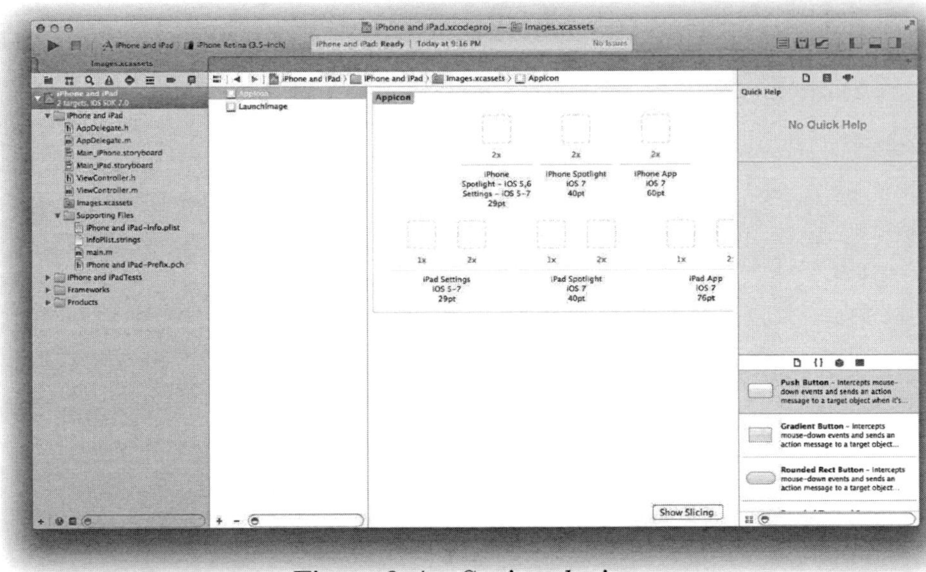

Figure 8.4 - Setting the icons

Once all of your icons are set, run the app to see if they took effect. If they don't appear, clean the project by clicking *Project » Clean*, then delete the app on the chosen device and run again. This time they will definitely appear!

8.5 Math, Math, Math

Without a doubt, you'll use math in programming. Whether you like it or not, math is now a part of your life (if it isn't already)!

In this section we'll go over the various math related functions that are not specific to Objective-C, but all programming languages in general. You'll use math to make games (probably Trigonometry, at least Geometry), but that's not what I'm going to teach. I'm just going to teach the basics of math in programming although it can later get pretty tricky with sine, cosine, and tangent. Luckily, the most difficult thing you'll learn here is modulus.

	Description	Example	Result
+	Addition	x = 4+2	x = 6
++	Add one to a variable	x++	x = x + 1
+=	Variable equals to variable plus another number	x += 5	x = x + 5
-	Subraction	x = 4-2	x = 4
- -	Subtract one from a variable	x- -	x = x - 1
-=	Variable equals to variable minus another number	x -= 5	x = x - 5

	Description	Example	Result
*	Multiplication	x = 4*2	x = 8
/	Division	x = 4/2	x = 2
sqrtf(x)	Square root of a float variable	x = sqrtf(4)	x = 2
abs(x)	Absolute value	x = abs(-5)	x = 5
powf(x, y)	Power of a float variable (exponent)	x = powf(3, 2)	x = 9
MIN(x, y)	Minimum of a range	x = MIN(2, 3)	x = 2
MAX(x, y)	Maximum of a range	x = MAX(2, 3)	x = 3
% (modulus)	The remainder of two numbers	x = 5%2	x = 1

This is pretty much it before you get into trigonometry. Most of this is pretty self explanatory, but I thought I'd share these functions with you as some functions require float variables as opposed to integers, and the fact the the minimum and maximum are capitalized for no particular reason.

Before you ever start getting into game programming, I recommend that you at least start geometry, if not trigonometry because games use many angle function and physics that you don't cover in school before geometry and trigonometry. Please hear me out on this! You will be tear your hair out attempting to make a game before properly learning about angles, sine, cosine, and tangent.

8.6 Opening and Previewing Files

Problem: You have a PDF file that you want to display in app upon the tap of a button.

Solution: Use the UIDocumentInteractionController to open and "interact" with a PDF file.

Sometimes programmatically displaying files can be tedious… this is one of those times. It's my personal belief that every programming language should have a simple, one line of code, way to open or display any type of major file. Sadly, this is not the case for most languages.

Let's jump into this by creating a new "Single View Application" project. Once created, go to the header file. In the header file, we must assign *ViewController* to the delegate of the document controller. To do this, add the following code after *@interface ViewController: UIViewController* :

```
<UIDocumentInteractionControllerDelegate>
```

Be sure to then make the header two sections like we've done throughout the book.

To finish off the header, initialize an action function. We'll use the action to open the PDF file. Once completed, head over to the method file:

```
-(IBAction)openPDF;
```

In the method file, add the action that we just declared. The first step in opening a PDF would be to locally initialize the document controller:

```
UIDocumentInteractionController *pdf =
[[UIDocumentInteractionController alloc] init];
```

Due to the fact that a different type of "controller" will be handling a file from our default file, we must assign a delegate:

```
pdf.delegate = self;
```

These next steps all involve determining the location of the PDF, and then setting the location to the document controller's assets. We first locally initialize a URL, and then set it equal to the location and file type extension of our PDF file (add a file to your project if you haven't already). Next, we have to set the controller equal to the URL, but oddly enough, we literally set *UIDocumentInteractionController* equal to the URL, not *pdf*, which comes right after this statement. This block of code is as such:

```
NSURL *pdfURL = [[NSBundle mainBundle]
URLForResource:@"FILE_NAME"
withExtension:@"pdf"];
    [UIDocumentInteractionController
interactionControllerWithURL:pdfURL];
    [pdf setURL:pdfURL];
```

We then end the action by properly presenting the file as a "preview". I don't know why it's called a "preview" as it displays the whole gosh darn file!

It may look like we're done, but there's still some more work to be done! Exit the action function. For some reason, *pdf.delegate = self;* is not good enough for the program to realize that our default controller is the delegate for the document controller. In order to set it straight, we have to add this (not in any function):

```
-(UIViewController
*)documentInteractionControllerViewControllerForP
review:(UIDocumentInteractionController
*)controller {

    return self;

}
```

The code will run smoothly now, but there are still some functions available to add additional functionality to your code. For example, you can add a function that runs code whenever the document controller is dismissed:

```
-(void)documentInteractionControllerDidEndPreview
:(UIDocumentInteractionController *)controller {

}
```

There are a whole bunch of other functions that you can use that cover pretty much every possible event. Feel free to mess with these functions as much as possible.

If you're done messing around with the code, go to the storyboard file, and then add and connect a button to the action function. When ready, build and run on your selected device.

8.7 Working With Instagram

Problem: You want to programmatically post an image to Instagram.

Solution: Edit the image size and extension, then use a URL scheme and document controller to post the image through the Instagram app.

Due to the fact that Instagram has yet to release an SDK, posting to their popular social network is not simple… at all. There are many steps to posting an image programmatically, so let's get started right away!

Once you have a new project ready, go to the header file. Since we'll be using the document controller, we must set the type of delegate:

```
<UIDocumentInteractionControllerDelegate>
```

If haven't already, add an image that you'd like to post (you don't actually have to post it). From here, declare an image in the first section of code:

```
UIImage *instaImage;
```

In the second section, declare an action function:

```
-(IBAction)post;
```

Now let's move onto the method file. In the method file, go to *viewDidLoad* and set the image we just defined to the image that you added earlier:

```
instaImage = [UIImage imageNamed:@"your.png"];
```

Initialize the action function so we can get started with the real meat of the code:

```
-(IBAction)post {

}
```

Since Instagram isn't a default app included with any iOS devices, we can't assume that the user has Instagram installed at any given moment. To check this, we must use Instagram's URL scheme inside of an if statement at the beginning of the function:

```
if ([[UIApplication sharedApplication]
canOpenURL:[NSURL
URLWithString:@"instagram://"]]) {

}
```

The first few things we need to do is to prepare to change the file extension of our image. This is done by accessing the app's home directory and creating a file with the extension ".igo":

```
NSString *documentsDirectory = [NSHomeDirectory()
stringByAppendingString:@"Documents"];
NSString *savedImagePath = [documentsDirectory
stringByAppendingString:@"Image.igo"];
```

The next step is to take the image we already added and render it to the standard Instagram picture dimensions while saving it as a new image. This is done using the *UIGraphics* imaging rendering code:

```
UIImage *newImage = instaImage;
    CGSize newSize = CGSizeMake(612, 612);
    UIGraphicsBeginImageContext(newSize);
    [newImage drawInRect:CGRectMake(0, 0,
newSize.width, newSize.height)];
newImage =
UIGraphicsGetImageFromCurrentImageContext();
UIGraphicsEndImageContext();
```

Since the document controller can interact with URLs, we must make our new image a data file, and then add the data to our path, *savedImagePath*. We can then create a URL that is set to our path which will change our image's extension to ".igo". The code looks as such:

```
NSData *imageData =
UIImagePNGRepresentation(newImage);
    [imageData writeToFile:savedImagePath
atomically:YES];
    NSURL *imageURL = [NSURL
fileURLWithPath:savedImagePath];
```

Finally we have everything set up for the document controller! We now need to locally declare a controller, set it's delegate, UTI (Uniform Type Identifier, not the gross one), URL, and annotation. Everything is probably self explanatory except for the UTI and annotation. A UTI (in programming) allows the app to determine what other apps on the device can open the bundle identifier set as the UTI. Annotations hold extra details about the information being sent. For example, the Instagram annotation holds the picture's caption as an object, and what type of data (a caption) as a key. The code is as follows:

```
UIDocumentInteractionController *doc =
[[UIDocumentInteractionController alloc] init];
    doc.delegate = self;
    doc.UTI = @"com.instagram.exclusivegram";
    [doc setURL:imageURL];
    doc.annotation = [NSDictionary
dictionaryWithObject:@"#iOSAppDevelopmentforTeens
byTeens" forKey:@"InstagramCaption"];
```

To finish this puppy off, we simply need to tell the document controller to present the apps capable of opening the UTI specified in our code at *CGRectZero*, or the bottom of the screen (0, 0):

```
[doc presentOpenInMenuFromRect:CGRectZero
inView:self.view animated:YES];
```

This is all code-wise! You can now go to Interface Builder, and drop and connect a button to the action function to top it all off! In case you're missing something, here's the code as a whole:

```
- (void)viewDidLoad
{

    instaImage = [UIImage
imageNamed:@"YOUR.png"];
    [super viewDidLoad];
}

-(IBAction)post {
    if ([[UIApplication sharedApplication]
canOpenURL:[NSURL
URLWithString:@"instagram://"]]) {

        NSString *documentsDirectory =
[NSHomeDirectory()
stringByAppendingString:@"Documents"];
    NSString *savedImagePath =
[documentsDirectory
stringByAppendingString:@"Image.igo"];

    UIImage *newImage = instaImage;
    CGSize newSize = CGSizeMake(612, 612);
    UIGraphicsBeginImageContext(newSize);
    [newImage drawInRect:CGRectMake(0, 0,
newSize.width, newSize.height)];
    newImage =
UIGraphicsGetImageFromCurrentImageContext();
    UIGraphicsEndImageContext();

    NSData *imageData =
UIImagePNGRepresentation(newImage);
    [imageData writeToFile:savedImagePath
atomically:YES];
    NSURL *imageURL = [NSURL
fileURLWithPath:savedImagePath];
    UIDocumentInteractionController *doc =
[[UIDocumentInteractionController alloc] init];
    doc.delegate = self;
    doc.UTI = @"com.instagram.exclusivegram";
    [doc setURL:imageURL];
```

```
    doc.annotation = [NSDictionary
dictionaryWithObject:@"#iOSAppDevelopmentforTeens
byTeens" forKey:@"InstagramCaption"];
    [doc presentOpenInMenuFromRect:CGRectZero
inView:self.view animated:YES];

    }

}
```

Ch. 8 Conclusion

I'm sad to say that the coding portion of the book has come to an end. The last chapter is all about getting your (finished) app on the App Store. I hope that you enjoyed this relaxed chapter. Even if you didn't read the whole chapter, I'm glad that you didn't take the easy way out and call it quits on Chapter 7. I truly enjoyed writing this chapter because I finally got the chance to teach someone my favorite programming concepts and they're able understand a lot of it!

Although you may not have read the whole chapter, you'll probably find yourself referring back to this section whenever you're a programming genius and are looking to perform some type of function in your next great app, and forgot how to code it. I know that personally, I did this constantly with my collection of programming books. Learning a programming language is just like (maybe more complicated than) learning a new speaking language, and it's okay that you have to refer to books to refresh your memory on a certain topic.

For this and the last chapter I will not be giving you a checklist in the conclusion due to the fact that the chapters jump all over the place. The last chapter is for those ready to submit their app to the

App Store, but feel free to read it now and refer to it whenever you're actually ready to take the last leap to app authorship!

Chapter 9

To Infinity and Beyond

The Final Steps to Getting Your App on the iOS App Store

So you've developed an app and want the whole world to see it. It's about time! As you probably know, the iOS App Store is one of the (if not) most popular app store on the Internet. It's understandable that you want to own a share of it! The only problem is how to not only submit your app, but get it approved.

Ever since Apple passed the 500,000 app mark, they've been getting more and more difficult with allowing average apps to get onto the App Store. Fear not though, the first section of this chapter will teach you what Apple doesn't tell you: how to get your app on the App Store when facing a rejection.

9.1 THE REJECTION LETTER - TIPS TO APP STORE APPROVAL

"What the flip! He's telling me how to cope with rejection yet I don't even know how to submit an app!?" Yes, yes I am. Truth be told my first app, all the way up to my latest apps, have been rejected before (thirty-two times). It's tough whenever an app gets rejected, especially when it's your first. All those hours and hours of work just to get a generic email vaguely stating why they "are unable to post this version" of your app. They state it without emotion as if you are some robot who doesn't care about their own hard work. Calling it

"Submission Feedback", they're just trying to cover up their real goal, rejecting apps. Whenever you get an app approved they don't give you any "feedback", they just tell you it'll be available within twenty-four hours. Those monsters.

Through my many experiences, I've developed a list of what not to do when facing rejection. I've shortened it to get rid of personal items and things that didn't really work. Here it is.

What Not To Do When Facing App Store Rejection

1. Do not respond with anything rude. This should be pretty self-explanatory. In the case that you have the same reviewer reviewing your app (very likely), they look back to your app's submission history and do not appreciate rude comments. They, like most adults, have to work a boring (sort of) nine-to-five job and probably feel your pain but just aren't allowed to show it (that's what I tell myself).

2. Do not submit an appeal to the Apple Review Board. The Apple Review Board is just made up of all of the reviewers, including the one that reviewed your app, so unless there is a very large misunderstanding from the person that reviewed your app, there is practically no chance that they will change the ruling. It's just another week long wait only to get rejected again. Personally, none of my appeals have ever been accepted.

3. Do not call the number provided with your Apple Review Board email. Yep, it's just someone who is representing your reviewer. They have even less sympathy for you, and just simply restate the rejection code provided in the initial email.

4. Do not submit the exact same build of an app again. They know. The reviewers all have a log containing information about the app each time you upload it. If you happen to get a different reviewer, they'll see the same notes and log of the last reviewer. If

you haven't made any changes at all to the app, then resubmitting immediately is just a waste of time. I'm guilty of doing this and can say that it's just another painful week of waiting.

5. Do not give up. Giving up on your hard work is just a way of letting Apple win. Put up a fight, give it your all! Listen to what Apple wants you to change and take it into consideration. If they're asking too much, try to do a little bit of what they ask and let them know that they're being unrealistic. Submit your app again and see what happens. In all seriousness, it's okay to give up on something every once in a while, but it's like cutting your wrists (not saying that's okay), you don't want to do it. When life (or Apple) throws you down, get back up and punch it in the face.

In my many experiences I've also come up with some very helpful tips for app submission.

What To Do When Submitting

1. Make sure that all of your account (and payment) details and app details are all in check. If you don't, you'll face instant rejection. Also make sure that any temporary text or screenshots are removed so that your app's page looks nice and clean.

2. Leave a note to the reviewer. In your app's metadata (details page), you have the opportunity to leave a note to your reviewer. Whether it's to clarify some app details or to brighten up his or her day, just leave a little something.

3. If rejected, listen. Whether you like it or not, you're stubborn, everyone's stubborn in some fashion. Apple just rejected your app and you don't want to make any changes to it. Please make

the next reviewers job easy and just listen to the suggestions in some way or another.

9.2 Creating Your Distribution Certificate

Remember that thing we made long ago called a "Development Certificate"? Well... we're about to do almost the exact same thing. It's fine if you don't remember because that gives this section even more purpose!

A distribution certificate is a certificate generated by Apple, using your credentials, that enables your developer account for App Store and Ad-Hoc (beta testing) distribution. Once you create and install

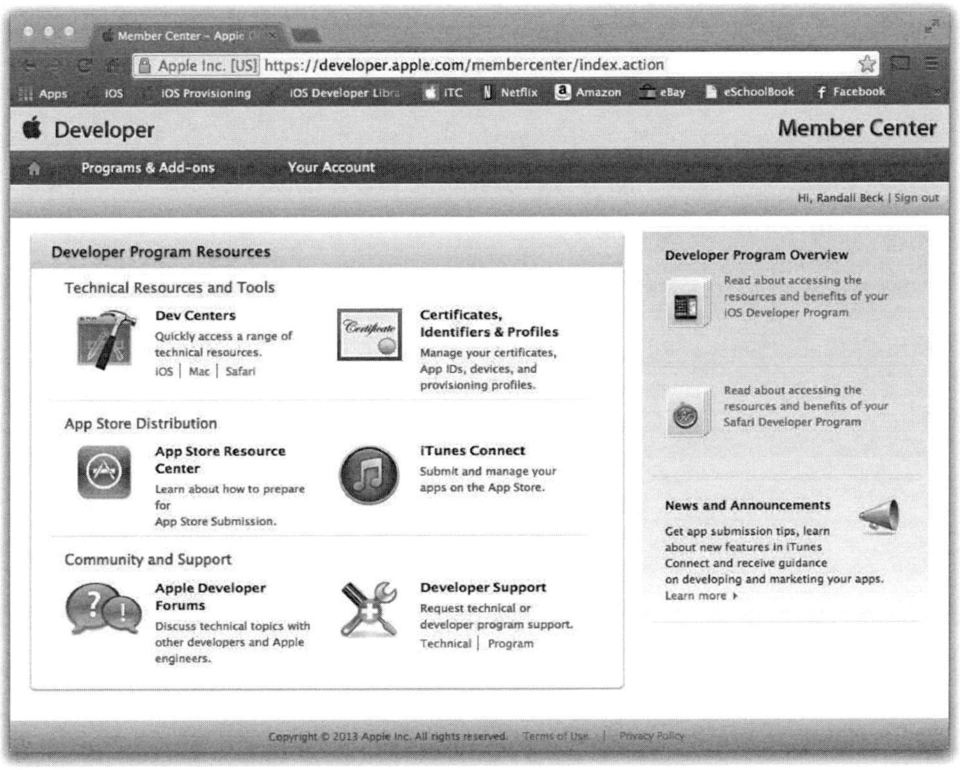

Figure 9.1 - Apple Developer Center

the certificate, you'll be able to create distribution provisioning profiles for your apps to use. Let's get distributin' like Rasputin (kinda rhymes)!

The first step is to simply find the place to get your certificate. This is just like before, visit the Apple Developer Member Center:

https://developer.apple.com/membercenter/index.action

From here, click on the link, "Certificate, Identifiers & Profiles". On that page, click "Certificates" under "iOS". To add a certificate, next to "iOS Certificates", click the plus button.

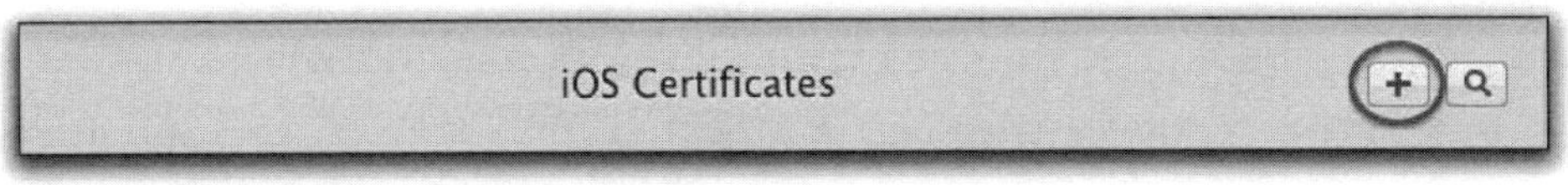

Figure 9.2 - Create a certificate

This page is a selection page to determine what kind of certificate you want. Since we already know what certificate we need, under "Production", select "App Store and Ad Hoc". Once completed, click "Continue".

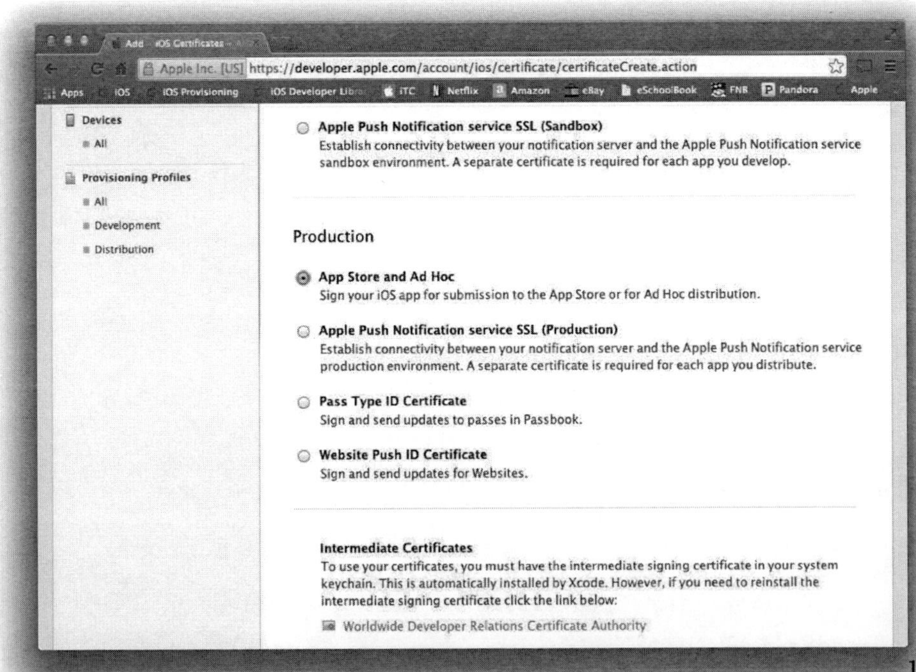

Figure 9.3 - Selecting a certificate

This next screen will ask you to manually generate a CSR file. Remember that file we created in Chapter Two? Thankfully, we can reuse that file (as long as one year or longer has passed since creation)! Hopefully you didn't misplace or delete the CSR file. If you did lose it, follow Apple's onscreen instructions or refer back to Chapter Two Section Four. Have it? Great! Push "Continue".

Now we need to upload our CSR file. Click "Choose File…" and find your CSR file containing the extension ".certSigningRequest". Once chosen press "Open", then "Generate". You'll be taken to the final screen in which you can download your distribution certificate. When finished downloading, click on the certificate in order for it to be installed on your computer under the Keychain Access app.

9.3 Creating Distribution Provisioning Files

Again, this is another lucky break… this step is very similar to creating a development provisioning profile. Honestly, the only difference is the type of use, one is for development, the other is for distribution. Similar to the development profile, you may use an Xcode wildcard to submit your app. The only time you cannot use a wildcard for distribution is when you use a special feature such as Game Center, iCloud, or iAd. The reason why they can't use a wildcard is due to the fact that Apple needs to store information about the app's usage, and the user needs to be able to access this information across multiple devices. If you were to use a wildcard, this information would not be able to be uniquely stored, and would be lost. In this tutorial, we're going to create an ID for your app and then a distribution provisioning profile.

Creating an App ID

Even if you don't plan on using any of the special features mentioned before that require a specific provisioning profile, you're still going to need an App ID. Each provisioning profile must be tied to an App ID in order to be created, and we still have yet to create an ID for the wildcard.

To create an ID, first visit the "Certificates, Identifiers & Profiles" section of the Member Center. Once you've reached that page, click "Identifiers" located under "iOS Apps". This page should look very similar to the "Certificates" visited earlier. Here we must also click the plus button next to "iOS App IDs".

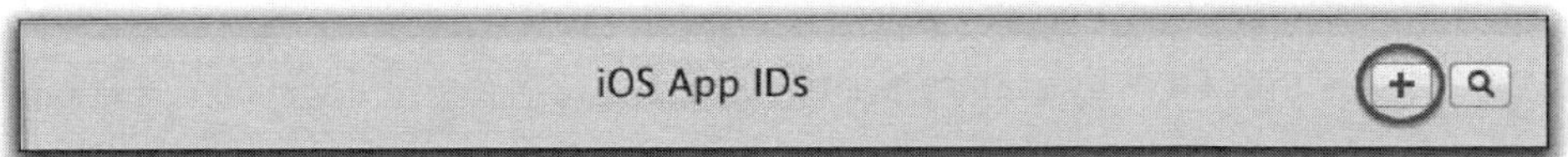

Figure 9.4 - Add an ID

On the ID creation page, the first field requires you to name the ID. Don't worry, this field is just for personal reference, so name it as you please. The next field, "App ID Prefix" represents who made the app. All of the choices will be some combination of numbers and letters, select the one with the text "Team ID" following the characters.

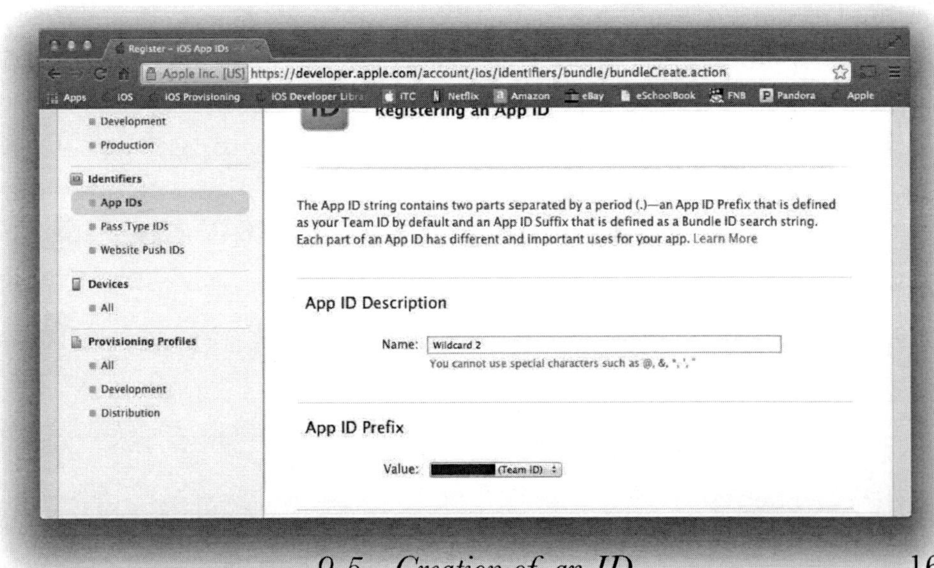

9.5 - Creation of an ID

Next is the suffix. This is where it gets really important. If you are planning on using "services such as Game Center, In-App Purchases, Data Protection, or iCloud", you must create an explicit App ID, or an ID that may only be used once. Otherwise, you may create a wildcard App ID which can be used over and over. If you are creating an explicit App ID, you MUST MUST MUST use the bundle identifier that you created your Xcode project with! Otherwise, the whole submission process will fail (example: *com.ch1pasoft.myNewApp)*! For the wildcard App ID, all you must type is the reverse domain followed by an asterisk (example: *com.ch1pasoft.*)*.

Once finished with that portion, scroll down to find an area to enable services that your app uses. This applies to those who are using an explicit App ID with some of the services stated on the webpage. These choices may be edited at any time once your ID has been created. If you are ready, click continue.

The final page just repeats everything that you selected on the last page. Review all of the items and then click "Submit" at the bottom of the page. Within a minute or so, your App ID will be ready for use!

Distribution Profiles

Now that we've got an App ID, we can finally create our distribution profile! From the "Certificates, Identifiers & Profiles" page, click on "Provisioning Profiles". On this page, select the plus button to get started with the creation.

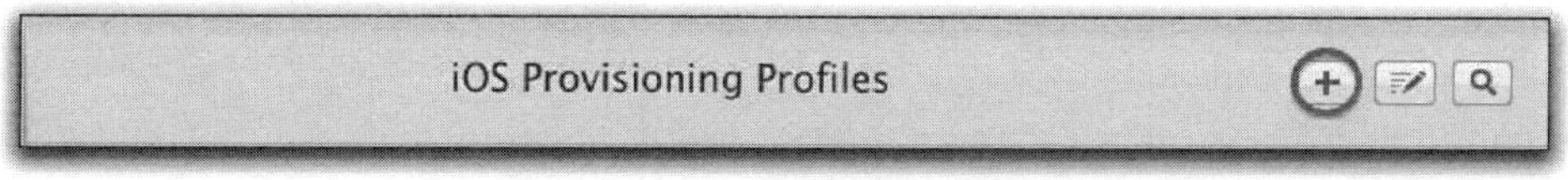

Figure 9.6 - Adding a profile

The profile page looks similar to the certificate creation page. The first step is to identify what type of profile we need. Since we're submitting the app to the App Store, we must select "App Store" under "Distribution" and click "Continue".

Now that we're on the configure page, we have to select which App ID we want to use. If you created an explicit ID earlier, select the name and identifier that exactly match your project's bundle identifier. If using the wildcard, select the wildcard ID you just made and hit "Continue". The next configuration page asks you to select your iOS Distribution certificate we created in the last section. Select it and press "Continue".

On the final page, we first have to name our profile. This is just for cosmetic and personal reference, so name it as you please. Once named, review the information on the page and then click "Generate". Once generated, download your new certificate and drag it to Xcode.

9.4 Adding an App to iTunes Connect

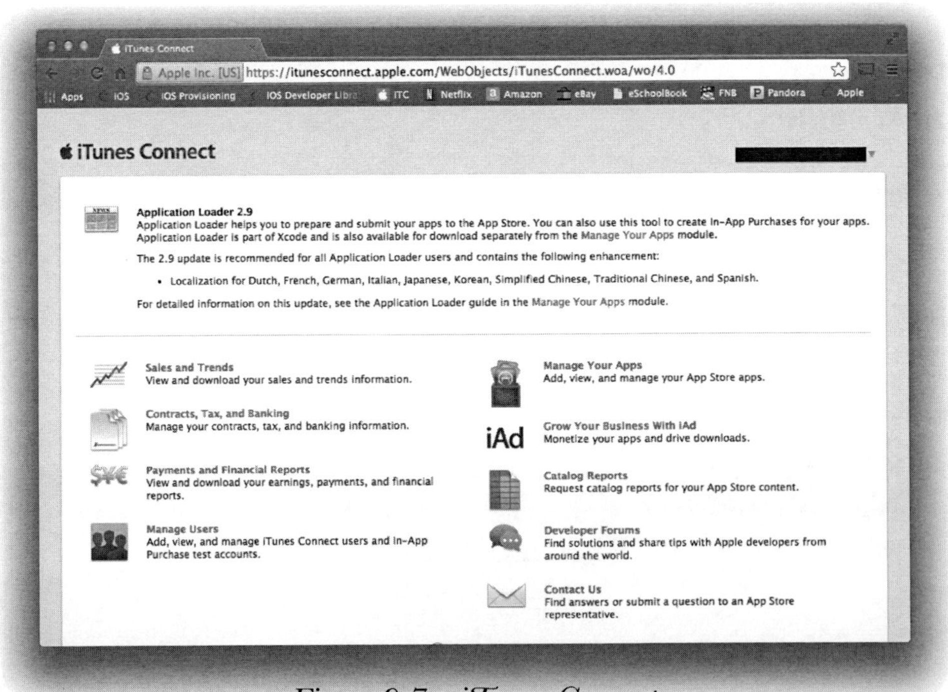

Figure 9.7 - iTunes Connect

iTunes Connect is the one spot where you manage everything App Store related. This includes viewing downloads, revenue, iAd information, managing payment information and users, signing contracts, and of course managing your apps. In this tutorial, we're just going to focus on the area where we can create a new app. Let's get started by visiting:

https://itunesconnect.apple.com/

Once you have reached the page and have logged in with your developer Apple ID, click the link titled "Manage Your Apps". The page that loads is where all of your apps are managed from. Since you probably don't have any apps on your account yet, the page will look pretty barren. To get started on inputing your app's information, click the "Add New App" button next to "Manage Your Apps".

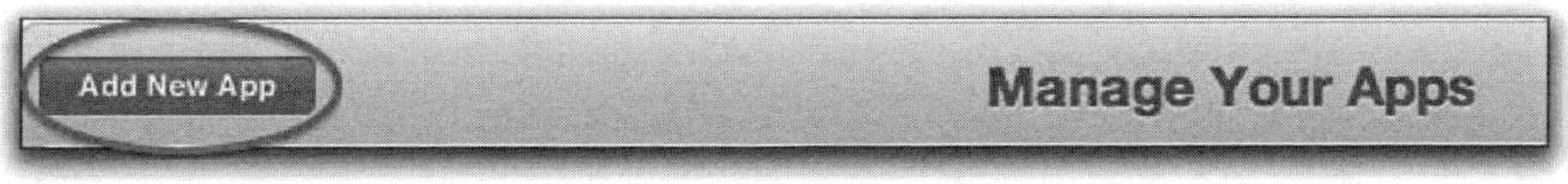

Figure 9.8 - Adding an app

You'll be taken to the first step of creating your app. The first page just lays down the basics of your app such as it's name (must have never been used before), the primary language, App ID

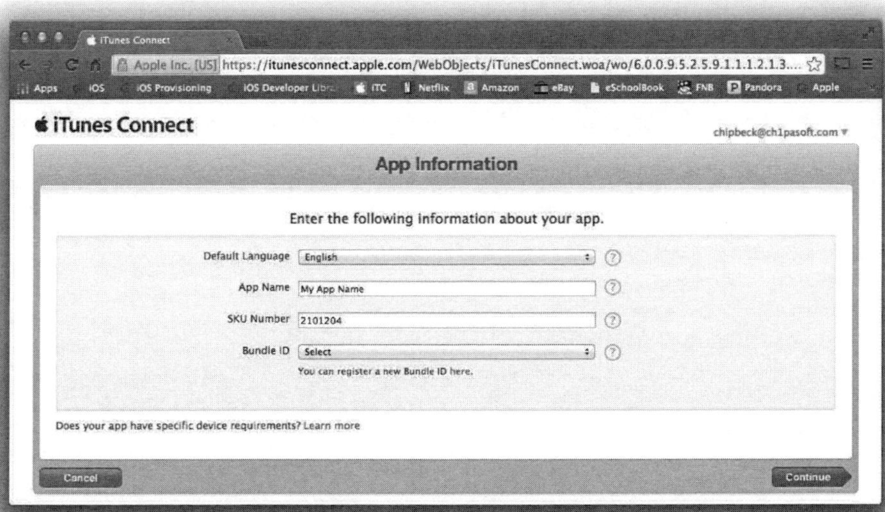

Figure 9.9 - App information page

(referred to as "bundle ID") and SKU. The SKU number is a product number that is just for personal reference, so don't worry there. The App ID is very important, so make sure it matches the provisioning profile you'll be building the app with. When ready, click "Continue".

The next section is pricing and availability. You can decide the price tier, and voluntarily enable educational discounts. The pricing is based on tiers due to the fact that the world doesn't run on a single currency (though free is always free). Apple has already decided the pricing based on tiers which you can view country by country on the pricing matrix that they provide on this webpage. Next let's look at the availability date. The date that you choose specifies the earliest that you'd ever want your app on the App Store. This doesn't dictate how fast your app will be reviewed though. If you set a date past the time that your app is approved, your app won't appear on the App Store until this date. If your date is before the approval date, the app

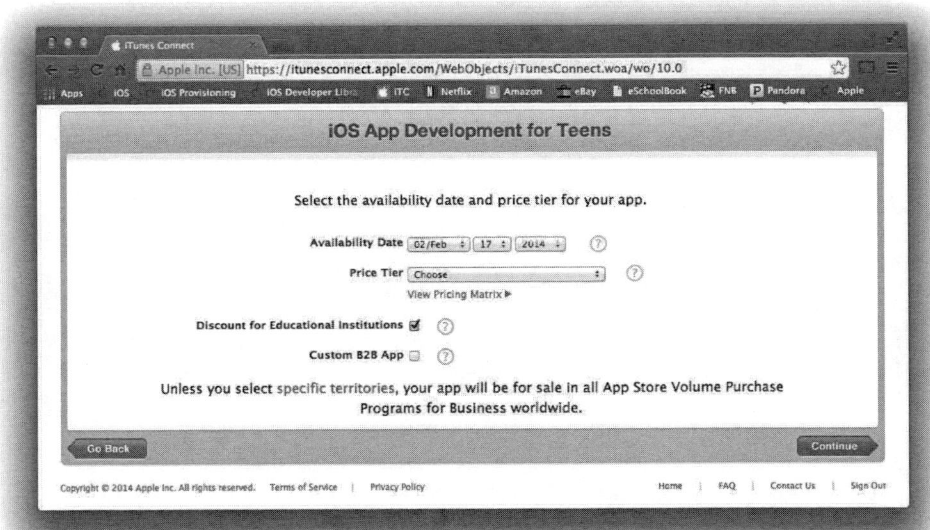

Figure 9.10 - Pricing and availability

will be available within twenty-four hours (or "immediately" as Apple calls it). Now for "Custom B2B App". A B2B or business to business app is something that is not available to the public, but just to people holding a business Apple ID. Only select this option if the app is not useful to the public, but meant to be purchased in bulk for businesses to distribute across all of their devices. Apps not under B2B may still be purchased by businesses, but not in the same way as B2B apps are purchased. When ready, click "Continue".

We've made it to the final webpage… it's a very long one! You'll fill out all of the details that relate to your app's App Store page here. Since there are too many details to simply summarize what you're going to be doing, let's jump into this!

The first section of the final webpage relates to some legal information, the version and copyright. The version number must match the version number specified in your Xcode project (default is "1.0"). Now to copyright. As soon as you create a unique file, it's automatically copyrighted in your name. That means that your app and it's contents are copyrighted to you at the moment of creation. Although, if your material appears to be someone else's copyrighted material, your copyright is nullified and you can face legal action (in short, use YOUR OWN material). Back to the webpage, for "Copyright" enter the year followed by your company name or legal name (example: © 2014 Randall A Beck II). To symbolize the copyright logo, you can type (c) or press *Option G.*

The next section is for setting your app's categories. You may use two categories, one primary and one secondary. The primary is the default category and should be the most relevant to your app. The secondary category is optional, but can be used to increase visibility.

Next is the rating. Apple's app rating process happens to be all set by the developer, not Apple themselves! This section is pretty self explanatory, just fill in each bubble that describes your app. The following ratings are possible: 4+, 9+, 12+, 17+. Note that some countries do not sell apps that fall into certain categories and ratings. Also, do not lie about your rating, Apple checks. Self-rating is just one way to make your review time shorter, don't lie and make it a longer wait! As an indirect feature for your rating and category, you can add your app to the "Made for Kids" section of the App Store. If your app is made for children, this features helps parents and children find your app.

Now we're up to the metadata. Metadata is information that describes data, a.k.a. your app. This section starts out with the description. This is the heart and soul of your app's App Store page. The selling point. Make sure you adequately describe your app, and remember that you can come back at any time to change this data. Next up are the keywords. This is how users will find your app. Unfortunately you're limited to only one hundred characters, so make the most out of your words. Now we have the website for app support. Make sure that the user is actually taken to your website, or a webpage where they can find support as Apple actually checks the URLs for usefulness. The URL for marketing is optional, but should be an ad for your app or your website. This link is featured at the bottom of your App Store page. As for the privacy policy, do not link one unless you actually have a privacy policy as this is covered by Apple when left blank.

Okay, the most rigorous section is past us now! We now have contact information and information for the reviewer. In the contact section, leave all accurate information about yourself so that your app's reviewer may easily contact you if need be. For the review notes, if you have anything you need to tell the reviewer, leave it here. I encourage you to leave a little note here for the reviewer even if you don't *need* to tell them anything as it might just help your review a little bit. If your app requires an account to use, make sure to leave a demo account username and password for your reviewer.

Do not change the EULA (End User License Agreement) unless you happen to be some big shot corporation with a lawyer writing the agreement! This is covered by Apple by default.

FINALLY, we have the "Uploads" section. This is where you'll upload your app icon, screenshots, and app area coverage (if you happen to have developed a mapping app). This is what deters some people as they don't have screenshots yet. Fear not, you can upload blank images of the correct dimensions now and change them later! Apple requires very specific dimensions, here is a chart that explains all of the acceptable dimensions and file formats:

	Description	Dimensions	Format(s)
Large App Icon	App icon used for advertising	1024x1024	.jpeg, .jpg, .tif, .tiff, .png
3.5-Inch Retina Screenshots	Screenshots for iPhone 4S and below	960x640, 960x600, 640x960, or 640x920	.jpeg, .jpg, .png
4-Inch Retina Screenshots	Screenshots for iPhone 5 and above	1136x640, 1136x600, 640x1136, or 640x1096	.jpeg, .jpg, .png
iPad Screenshots	Screenshots for iPad	1024x768, 1024x748, 768x1024, 768x1004, 2048x1536, 2048x1496, 1536x2048, or 1536x2008	.jpeg, .jpg, .tif, .tiff, .png

	Description	Dimensions	Format(s)
Routing App Coverage (Optional)	Coverage file that specifies what geographic region an app covers	N/A	.geojson

Hopefully you've passed through the gauntlet successfully and are ready to save your project. Whenever ready, press "Save". You'll be able to come back at any time to edit this information (some items can still be edited once published on the App Store).

Now you have a full iTunes Connect page! From here you can review your app and the information that you filled out. Whenever ready, click on your app from "Manage Your Apps" and click the button "Ready to Upload Binary" from the top right corner next to your app's name.

Figure 9.11 - Are you ready to upload!?

Once clicked, you'll be asked a couple questions that Apple needs to know before the submission of each app. When finished, you'll be able to finally submit your app for review!

9.5 SUBMITTING YOUR APP

The time has finally come to actually upload your app to Apple's servers for review! You're on the final stretch, don't give up now! All we have now is to build and submit your app. Let's get started!

Open up your project in Xcode. Go to the project's main section which contains all of your app's main information (identified as a "target"). On this page check to make sure that your app's version number is the same as you entered on iTunes Connect. The next step is to open the tab named "Build Settings". From here, scroll down to the section titled "Code Signing". You'll see the setting "Code Signing Identity". In this area, change the settings "Debug" and "Release" to your app's distribution profile. If your distribution

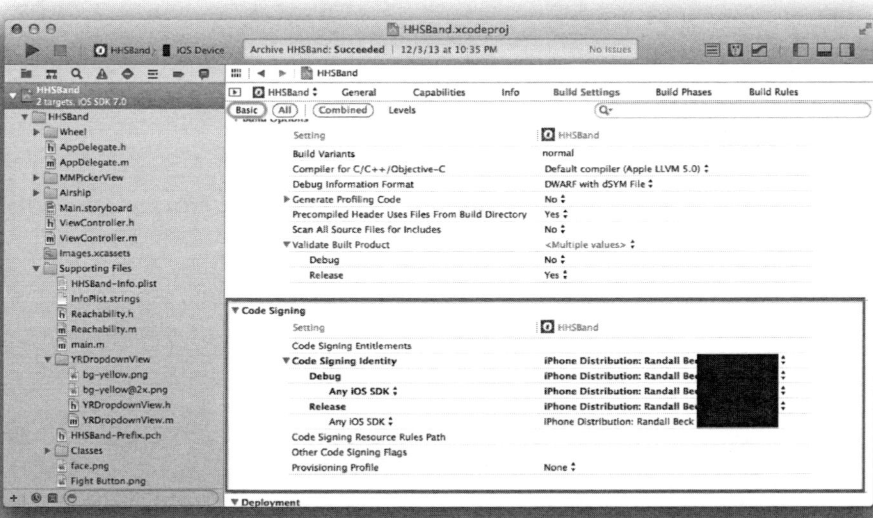

Figure 9.12 - Code signing

profile doesn't appear, simply click on the actual file on your computer and it should be added to Xcode. Once everything has been set to distribution, clean your project *Product » Clean* or ⇧⌘K. Next, build your project *Product » Build* or ⌘B. After you've cleaned and built your project, archive your project *Product » Archive*. Once your app has been archived, Xcode Organizer will be opened under the "Archives" tab. When ready, hit the button that says "Distribute…"!

Figure 9.13 - Xcode Organizer archive section

Once you press the button, you'll be asked to login to your Apple ID again, and you'll choose which app you'd like to submit (even though there is only one choice). The upload will start and everything should go swimmingly if you followed all of the steps correctly! CONGRATULATIONS on submitting your first app!

9.6 Where to Go from Here

> "Let all the women and fathers of children go from here." - Enjolras (Les Misérables)

Only if you could see my face to see how proud I am that you have finally reached the conclusion to *iOS App Development for Teens by Teens: Beginning iOS App Development*! By finishing the book, you are now an intermediate level app developer! I'm sorry, but I can't hold your hand forever so we need to talk about where to go from here.

There are so many options and possibilities now that you have the ability to develop apps from scratch. Don't reminisce on your past, think of all of your future possibilities. You now possess a skill that not many other pre-twenties people know how to do, and that gives you purpose. Purpose to go out and find a way to make the world a better place for all people, as well as future generations. I look at myself and can feel a sense of accomplishment. Two years ago I never saw myself writing a book or having a Kickstarter project, but life evolves unexpectedly, one day you may be figuratively (or literally) stuck in a ditch. It's easy to give up hope, but just think of what tomorrow could bring, and life becomes amazing and beautiful.

I do hope that you enjoyed your experience as I know I immensely enjoyed my experience writing this! Below is a list of future possibilities and helpful tips I've learned from experience.

Possibilities

- Do what you love most. Develop an app about something that you truly care about. It will make programming a little less painful.
- Contact local businesses. Spread the word that you can develop apps. Local businesses love supporting young people who have taken charge of their life. This helps you show off your skills even more.
- Keep learning. No matter how much you know about app development, there will always be more stuff to learn. Apple keeps updating their code and third-party developers always release new SDKs available for use. There will never be a dull moment in your journey.
- Stay up all night and write the next big app hit!
- Go to sleep because programming is truly hard work!

Tips

- If you ever get stuck, ask for help. Never be afraid to ask others for help. Although there are some ignorant people on the Internet who believe it is their job to make you feel stupid, there are also plenty of helpful ones who love helping others succeed. For programming, there is a great community website, StackOverflow.com, which is an amazing resource for any type of programming question. Another great resource that happens to be Apple-only is the Apple Developer Forums, a place where developers speak to other developers to help each other out.

- Be humble. Do not try to exaggerate your skill set, it only leads to embarrassing consequences. If you try to tell a client that you can do something that you actually can't, it's possible to burn bridges and finances.

- Try and try again. Apple can be cruel. Life can be cruel. If you hit some bug that seems impossible to solve or if your app is rejected by Apple or a client, try again. Trying over and over can't hurt, you have nothing to lose. You only lose whenever you give up. I've repeated this theme throughout the book on purpose, I gave up once and it led to resentment. I don't want you to regret anything you do in your life, and giving up only leads to regret. Just think of the lives you can change. Even if you just change one life for the better, isn't life worth living?

Thank you all,

Chip Beck

Hermitage, PA

Fifteen